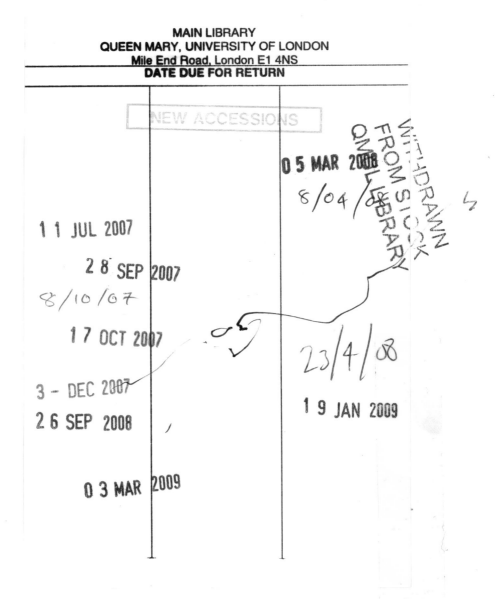

Antennas and Propagation for Body-Centric Wireless Communications

For a listing of recent titles in the *Artech House Antennas and Propagation Library*, turn to the back of this book.

Antennas and Propagation for Body-Centric Wireless Communications

Peter S. Hall
Yang Hao

Editors

**ARTECH
HOUSE**

BOSTON | LONDON
artechhouse.com

Library of Congress Cataloging-in-Publication Data
A catalog record for this book is available from the U.S. Library of Congress.

British Library Cataloguing in Publication Data
Antennas and propagation for body-centric wireless communications.—(Artech House antennas and propagation library)
1. Antennas (Electronics)—Design and construction
2. Radio wave propagation   3. Wireless communication systems
4. Human engineering
I. Hall, Peter S.   II. Hao, Yang
621.3'824

ISBN-10: 1-58053-493-7
ISBN-13: 978-1-58053-493-2

Cover design by Igor Valdman

© 2006 ARTECH HOUSE, INC.
685 Canton Street
Norwood, MA 02062

10 9 8 7 6 5 4 3 2 1

Contents

Preface

Body-centric wireless communications systems (BWCS) will be a focal point for future communications from an end-to-end user's perspective. It will play a key role in the development of fourth generation (4G) mobile communications, making convergence and personalization happen. Wireless communications is making inroads into every aspect of human life, and it is becoming an integral part of the human body communication system (HBCS). Advancements in wearable hardware, embedded software, digital signal processing, and biomedical engineering have made it practically possible for human-to-human networking with the use of wearable sensor communications.

We can say that BWCS is paving a path towards 4G communications. Two points are extremely important in achieving this, namely, the challenges and the definition of 4G. Let us first summarize the challenges.

1. Application-driven challenges:
   - Data fusion (aggregate and filter);
   - Support of multiple data rates;
   - Robustness, zero maintenance;
   - Security and privacy at low energy cost;
   - Localization.

2. Networking challenges:
   - Multihop and its implications;
   - Dynamic topology.

3. Technology challenges:
   - Low-complexity/low-power designs;
   - Smart personal networks and sensors;
   - Integration of heterogeneous networks considering BWCS.

This book focuses on the BWCS possibilities, technology, and problems. Special descriptions of the electromagnetic properties of the body, as well as the on-body communication channels at microwave frequency bands and at low frequency bands, are given in the book. Perspectives for body-centric ultrawideband (UWB) communications are described, and techniques for wearable antennas for cellular and wireless local area network (WLAN) communications and for body-sensor networks are introduced. The special circumstances for antennas, propagation for telemedicine, and for wireless implants are also presented. BWCS can be immediately affiliated to many different wireless technologies, and the possibilities presented by those technologies are many. Nevertheless, the possibilities and the advantages of affiliation to the future fourth generation communications technologies will be much larger, and the perspectives for BWCS-4G are therefore numerous.

A very important subject related to 4G is convergence. Convergence can be interpreted in several ways—namely, convergence of wireless and wired networks; convergence of communications, consumer electronics, and computer technology; and convergence of services. These are illustrated in Figure P.1.

From the perspective of 4G terminals, one can also see a clear trend in which several functions and capabilities converge into a single terminal, which will thus be a multimode (multistandard) multifunction terminal. Convergence will also take place at the network level, where several homogeneous networks will appear to be merged into a single "network of networks," as shown in Figure P.2.

Wireless body area networks (WBANs), wireless sensor networks (WSNs), and personal area networks (WPANs) are all emerging technologies that have a wide range of potential applications, such as in health care, smart home, surveillance, monitoring, mobile entertainment, and so forth.

Antennas and propagation are the most basic points for integrating WBANs, WSNs, and WPANs into future wireless heterogeneous networks, which is a necessary step to shape the 4G landscape. For obvious reasons, body-centric communication is becoming a very attractive research area for both academics and industrialists. Body-centric communications is a research topic combining WBANs, WSNs, and WPANs, as illustrated in Figure P.3.

This is the first book to give a broad treatment to antennas and propagation for body-centric wireless communications, with the objective of having a single-source

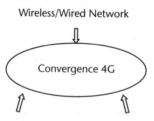

Figure P.1   Multiple convergence in 4G.

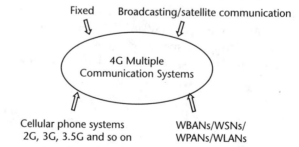

Figure P.2   Network convergence in 4G.

Figure P.3   WBANs, WSANs, and WPANs.

reference that offers in-depth coverage written by top experts. The book is suitable to young researchers, lecturers, practicing engineers, and technocrats. It is based on the contributions of several prominent researchers around the world who are involved in the development of the different areas within BWCS. The book will be an important tool for professionals working on the application of wireless communications in the field of biomedical engineering, biosensor networking, electromagnetic characteristics of the human body, and human-to-human networking.

Ramjee Prasad
Flemming Bjerge Frederiksen
Aalborg, Denmark
August 2006

CHAPTER 1

Introduction to Body-Centric Wireless Communications

Peter S. Hall and Yang Hao

1.1 What Are Body-Centric Communications Systems?

The ever-growing miniaturization of electronic devices, combined with recent developments in wearable computer technology, are leading to the creation of a wide range of devices that can be carried by users in their pockets, or otherwise attached to their bodies [1–3]. This can be seen as a continuation of a trend spearheaded by the mobile phone, which over the last decades has become smaller and more convenient for personalized operation. Alongside this trend, there have been a number of body-centric communication systems for specialized occupations, such as paramedics and firefighters, as well as continuing interest for military personnel. The development of the mobile phone can be characterized from the user perspective by three phases. First, equipment was large and heavy, and used by those people whose job required it. Then, the business community saw it as a way of improving business operation. Finally, the mobile phone became popular with the general population, who used it for social and entertainment purposes, and more recently as a fashion accessory. It is quite possible that wearable computers will follow the same path. Figure 1.1 shows a wearable computer developed at the University of Birmingham [4], as a test bed for a wide range of studies into hardware and software architectures and user applications. It is obviously bulky and inconvenient to use, and has wired interconnections. However, further miniaturization is expected, and it is clearly desirable to remove the wired interconnections. It is also likely that power will be provided at each body unit, and that data transfer will use high-capacity wireless communications [5]. Other current uses of wearable computers include warehouse operators and garage mechanics. Figure 1.2 shows the wearable computer as a fashion accessory.

Body-worn equipment is also used for health monitoring. Figure 1.3 shows the Senswear Pro2 Armband, a metabolic, physical activity, and lifestyle halter [6]. It is worn on the triceps of the right arm for up to two weeks continuously, and allows calculation of energy expenditure and quantification of metabolic physical activity. Physiological body signals from five sensors (two accelerometers, skin temperature, near body temperature/heat flux, and galvanic skin resistance) are used, in combi-

(a)

(b)

Figure 1.1 University of Birmingham wearable computer: (a) headset with video display, microphone, and earpiece; and (b) miniaturized PC.

nation with free-living activity recognition patterns, to calculate energy consumption based on specific algorithms.

Medical implants for monitoring, diagnosis, and activity have been studied for some time. The opportunities created by nanotechnology and microtechnology now open up the possibility of much more widespread use and application, for the wider

Figure 1.2   Wearable computer as a fashion accessory.

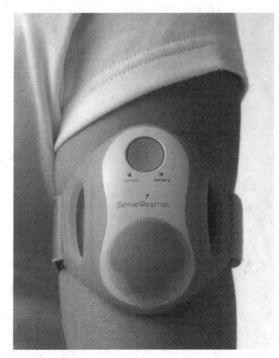

Figure 1.3   Senswear Pro2 Armband—Metabolic, physical activity, and lifestyle halter. (Courtesy of Apc Cardiovascular U.K.)

benefit of society. In addition to sensors and drug delivery mechanisms, communications is a vital part of this implementation process, both for monitoring of internal body conditions and for signaling actions to be taken by the implant. These implants might also be used in conjunction with a body area network (BAN), as shown in Figure 1.4 [7]. Optimum designs of such BANs and implants means that a full understanding of antennas and propagation into and through the body is needed.

The preface to this book has given a perspective based on fourth (and beyond) generation mobile and personal communications. There, body-centric communications takes its place firmly within the sphere of personal area networks (PANs) and body area networks (BANs). The content of a PAN or BAN involves a range of body-centric communications needs and requirements. These can be classified as:

- Communications from off-body to an on-body device or system—*off-body*;
- Communications within on-body networks and wearable systems—*on-body*;
- Communications to medical implants and sensor networks—*in-body*.

The italics show the shorthand nomenclature that we use in overviewing the book in Section 1.5. This nomenclature implies a partitioning of the PAN and BAN space into three areas. The first is where most of the channel is off the body and in the surrounding space, and where only one antenna in the communications link is on the body, which we call the off-body domain. The second is where most of the channel is on the surface of the body, and both antennas will be on the body, called the on-body domain. The last is where a significant part of the channel is inside the body and implanted transceivers are used, and we call this the in-body domain. While this is not a perfect subdivision, it does serve to highlight some of the different challenges for antennas and propagation in the body-centric system.

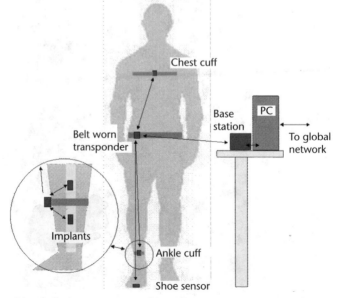

Figure 1.4 Wearable medical support network.

1.2   Off-Body to On-Body Communications

Communications from localized base stations or broadcast stations to transceivers located on the body have been studied extensively [8–10]. In the mobile phone area, such investigations include studies of the propagation characteristics of urban and rural environments, and the performance of body-worn antennas with variations of body proximity and orientation. There are many publications on this topic, and we have included little of it in this book, except those topics that relate to the use of ultrawideband techniques in Chapter 5, which is a relatively new research area. However, we have noticed that the topic of fabric-based antennas, discussed in Chapter 6, is now gaining prominence, since these antennas can significantly increase the performance of body-worn equipment in communications to local base stations. Such antennas need to orient the radiation pattern away from the body, while simultaneously providing all-round coverage. In addition, it is important to screen the antenna from the body, to prevent the body tissue from degrading the antenna efficiency. This can be achieved by the use of much larger ground planes than are possible in miniaturized equipment. The challenge for a fabric-based antenna is to maintain good performance in the face of changes in the body posture, yet still be unobtrusive.

1.3   On-Body Communications

There are very few examples of wearable computers in general use today, although this is likely to change in the near future. Figure 1.1 shows a serious drawback, in that many demonstrator systems are presently wired. This is undesirable, due to reliability issues surrounding constantly flexing cables and connectors, the weight of such cables, and the sheer inconvenience to the user. A number of other connection methods have been proposed for this purpose, including smart textiles and communication by the currents in the user's body. Each of these methods has their own advantages and drawbacks. Among the drawbacks of the smart clothes, for example, is the need for a special garment to be worn, which may conflict with the user's personal preferences. Similarly, body current communication is limited because it has a relatively low capacity. For real-time video transfer around the body, very high data rates will be required.

Wireless radio connectivity is an obvious option for connecting body-worn devices. Several standards for wireless connections between small, closely spaced devices have been developed, including Bluetooth, BodyLAN, and Zigbee. These types of connections can provide high levels of flexibility and comfort to the user, and therefore have received much attention. There are three primary criteria for wireless modules for on-body communications. First, they must support the high data rates expected in the future. Second, they must be small and lightweight. Both of these suggest the use of high frequencies. Third, they must consume the minimum of power, which implies highly efficient links. In terms of antennas and propagation, efficient design requires both good understanding of the properties of the propagation channel involved, and the development of optimized antennas.

1.4   Medical Implants and Sensor Networks

In recent years, several exciting developments, such as submicron electronics, nanotechnology, and microelectromechanical systems (MEMS), have emerged, all of which will have a profound effect on medicine. These technologies will allow the construction of highly intelligent, microscopic, implantable sensors; mobile robots; and drug release devices. These devices will perform in vivo diagnostic and therapeutic intervention, improving the quality of life for many patients. There are now a number of conditions where implants are used to improve patient lifestyle, such as heart pacemakers [11], and cochlea implants [12]. There are also significant future aspirations, such as the projects within the EC Integrated Project, Healthy Aims [13], which include pressure sensing in the brain cavity, glaucoma sensors, and retinal implants. Other future aims are the lab-on-a-chip for internal diagnosis or automatic drug delivery [14], and interfacing to nerve endings for communications to the brain [15]. As an example of the significance of such developments, there are patients with severe spinal injuries who are fully aware but are unable to communicate, since they have no muscular control. Implanted sensors connected to the spinal cord, or even the brain, could give back the ability to communicate, providing an almost immeasurable impact upon the quality of life for these patients.

There are, however, many problems that must be overcome if this implant technology is to be widely exploited and used commercially. There is now much work to develop new sensor technology, much of which is being enabled by nanosystem and microsystem research [13, 14]. However, there is much less work being done on other important aspects of the implant system, namely, communications. This will have a crucial effect on the practicability of future implant systems, where the potentially very small size of the sensor needs to be matched by miniaturization in the communication technology. Electromagnetics is one of the enabling technologies that has to be applied within the framework of communications networks and user needs. There is also the concept of a hierarchy of implants applied to the head, as shown in Figure 1.5. The higher levels, some of which may be outside, or on the body, have high intelligence, and thus have high communication rate needs; those at

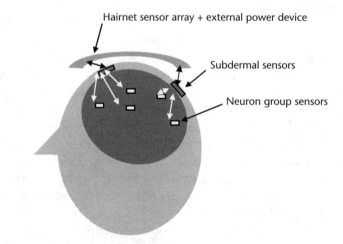

Figure 1.5   Communications to medical implants.

the lower levels become smaller and have smaller communication requirements. The subdermal sensors could read body data as well as being relays to the single, or group, neuron (or other body data) sensors, which are ideally at the submicron scale. At this size, sensor noise, either electrical or mechanical, will limit performance. The larger implants may be active (i.e., with internal power sources), and the smaller ones may be passive. In either case, communications is important. Sensors are now becoming available for all levels. The feasibility of providing communications will be limited both by the fundamentals of electromagnetics, and the antenna and transceiver technology needed to implement it.

1.5 Layout of the Book

The chapters and authors in this book have been selected to give overviews of work in the three areas of body-centric communications, as noted in Section 1.1. A guide to their content now follows. It can be read in conjunction with Figure 1.6, which shows the relationship of the chapters to each other and to the topic of personal and body area networks. The figure shows that Chapter 2, in addition to the Introduction and Summary, are relevant to all parts of body-centric systems. Only Chapters 3, 4, and 9 deal with specific domains, while the remaining chapters overlap areas between domains.

Chapter 2 discusses the electromagnetic properties of the body, including the material properties. Radiowave propagation models and specifications of antenna performance for an on-body or in-body antenna are introduced. The chapter also presents an introduction to numerical modeling techniques for antennas and propagation, and discusses body phantoms.

Chapter 3 gives an overview of on-body channels at microwave frequency bands, and gives a brief review of applications of microwave on-body communications, from current handset-headset communications to future applications, such as support for medical sensor networks, emergency personnel, and personal entertainment. It gives information about on-body channel measurements, channel characterization and modeling, antenna and system design, and link budget analysis.

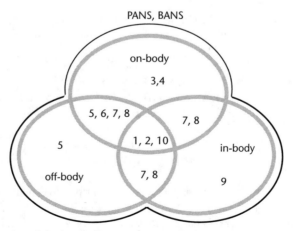

Figure 1.6   Interaction of the book chapters.

Chapter 4 deals with on-body wireless communications at low-frequency bands, and gives an overview of on-body channel performance at these frequencies. The chapter presents equivalent circuit models, numerical analysis, experiments using human phantoms, transmission mechanism, and overviews various applications.

Chapter 5 gives an overview of UWB as an enabling technology for body-centric communications. UWB antennas are discussed and analyzed. UWB channel measurement and simulation methodology, as well as channel characterization and modeling for PANs and BANs, are presented.

Chapter 6 introduces wearable antennas for cellular and WLAN communications, and is aimed primarily at communications from on to off the body. The chapter presents different antenna solutions for these communications methods and smart antenna fabrics.

Chapter 7 deals with body-sensor networks, including the basics of biomedical radio frequency (RF) telemetry and biosensor systems. Antenna designs for body sensors and different applications in military and space are presented.

Chapter 8 gives an overview of antennas and propagation for telemedicine. Various telemedicine applications are presented, and some principles of numerical modeling of the human body for telemedicine, together with system design principles, are introduced.

Chapter 9 introduces antennas and propagation for wireless implants. The chapter deals with RF biotelemetry, and focuses especially on low-profile antennas for implantable medical devices, and at antennas that can operate successfully in a lossy dispersive medium.

Chapter 10 gives an overview of the book, and includes tables showing in which chapter various communications standards and applications are considered. Conclusions are then given on the state of the art of body-centric wireless communications. Finally, some future challenges are presented.

References

[1]   Baber, C., et al., "Ergonomics of Wearable Computers," *Mobile Networks and Applications,* Vol. 4, 1999, pp. 15–21.

[2]   Shivers, O., "BodyTalk and the BodyNet: A Personal Information Infrastructure, Personal Information Architecture," Note 1, MIT Laboratory for Computer Science, Cambridge, MA, December 1, 1993.

[3]   Zimmerman, T., "Personal Area Networks: Near Field Intra-Body Communications," MIT Media Lab, *IBM Systems J.,* Vol. 35, No. 3 & 4, 1996.

[4]   Baber, C., et al., in M. A. Sasse and C. Johnson, (eds.), "A Wearable Computer for Paramedics: Studies in Model-Based, User-Centred and Industrial Design," *Interact'99,* Edinburgh, U.K., August 30–September 3, 1999, pp. 126–132.

[5]   Carvey, P. P., "Technology for Wireless Interconnection of Wearable Personal Electronic Accessories," *IEEE Workshop on VLSI Signal Processing, IX,* October 30, 1996, pp. 13–22.

[6]   Bodymedia Body Monitoring Technologies, http://www.bodymedia.com.

[7]   Spratley, J. P. F., et al., "Communications for In-Body Microsensor Applications," *Int. Workshop on Wearable and Implantable Body Area Networks,* Media Lab, MIT, April 2006.

[8]   Famolari, D., and P. Agrawal, "Architecture and Performance of an Embedded IP Bluetooth Personal Area Network," *2000 IEEE Int. Conference on Personal Wireless Communications*, December 17–20, 2000, pp. 75–79.

[9]   Stratis, G., et al., "Composite Antenna Pattern for Realistic Ray Tracing Simulations," *IEEE Antennas and Propagation Society Int. Symp.*, 2003.

[10] Schwendener, R., "Indoor Radio Channel Model for Protocol Evaluation of Wireless Personal Area Networks," *13th IEEE Int. Symp. Personal, Indoor and Mobile Radio Communications*, Vol. 2, September 15–18, 2002, pp. 891–895.

[11] Perrins, J., "Types of Sensor for Rate Responsive Pacemakers," *IEE Colloq. on Int. Cardiac Implants*, January 29, 1993.

[12] Loizou, P. C., "Introduction to Cochlear Implants," *IEEE Signal Processing Magazine*, September 1998, pp. 101–130.

[13] http://www.healthyaims.org.

[14] Figeys, D., "Lab-on-a-Chip: A Revolution in Bio. and Med. Sciences," *Anal. Chem.*, Vol. 72, No. 9, May 1, 2000, pp. 330A–335A.

[15] Mingui, S., et al., "Data Communication Between Brain Implants and Computer," *IEEE Trans. on Neural Systems and Rehabilitation Eng.*, Vol. 11, No. 2, June 2003.

CHAPTER 2

Electromagnetic Properties and Modeling of the Human Body

Peter S. Hall, Yang Hao, Hiroki Kawai, and Koichi Ito

Body-centric communications involves the interaction of electromagnetic waves with the body. In studying this interaction, it is most important to understand the electromagnetic properties of body tissues. These properties vary significantly with tissue type and frequency. To enable the development of antennas and transceivers for such communications systems, modeling of the body is a necessary task. The body can be modeled either with physical phantoms, made from solid, liquid, or gel materials; or with numerical phantoms embedded in numerical electromagnetic codes. The tissue properties and modeling techniques are addressed in this chapter.

2.1 Electromagnetic Characteristics of Human Tissues

The dielectric properties of body tissues at radio frequencies and microwave frequencies have been examined by several authors. One of the first extensive reviews and tabulations of previous results was made by Durney et al. [1]. Other early reports summarized specific resistance of tissues [2] and tabulated the dielectric properties of tissues in the frequency range from 10 kHz to 10 GHz [3]. An historical perspective was given in [4], which was extended to include more recent data (to 1990) in [5].

Use these studies to verify experimental data.

A recent comprehensive study is described in [6]. One of the most difficult problems associated with the determination of tissue parameters is that of obtaining samples. It is not possible to make measurements on live tissue, due to the need to have the measuring device in contact with the tissue. For this reason, all studies use dead tissue. In [6], excised animal tissue, mostly ovine, from freshly killed sheep, along with human autopsy materials were used. In addition, human skin and tongue in vivo was measured. An open-ended coaxial probe technique [7] was used for all measurements. The animal tissue was measured within two hours of death, while the autopsy material was measured within 48 hours. Measurements were made over a frequency range from 10 Hz to 20 GHz. Since the dielectric parameters vary significantly with frequency, a model based on the summation of 4-Cole-Cole expressions is used [6],

in vivo in death

$$\varepsilon(\omega) = \varepsilon_\infty + \sum_{m=1}^{4} \frac{\Delta\varepsilon_m}{1 + (j\omega\tau_m)^{(1-\alpha_m)}} + \frac{\sigma_j}{j\omega\varepsilon_0} \qquad (2.1)$$

where ε_∞ is the material permittivity at terahertz frequency; ε_0 is the free-space permittivity; σ_j is the ionic conductivity; and ε_m, τ_m, and α_m are material parameters for each dispersion region.

Table 2.1 gives the various parameters needed to find $\varepsilon(\omega)$ from (2.1), at any frequency, for a range of body tissue types [8]. Table 2.2 shows conductivity, relative permittivity, loss tangent, and penetration depth for these tissues at a frequency of 2.5 GHz, derived using (2.1) and Table 2.1. This frequency is chosen here to correspond to the measurement frequency used in Chapter 3 for on-body propagation channels, and is close to that used in Chapters 5 and 6, which use low microwave frequencies. Other chapters use different frequencies. Chapter 4 uses 10 MHz, and Chapter 8 uses a range of frequencies, including those around 400 MHz and 2.45 GHz. Chapter 9 also examines the use of 403.5 MHz and 2.45 GHz, and gives a useful comparison of some tissue properties at these two frequencies. There are significant differences in relative permittivity and conductivity across a wide frequency range, as is shown in Figure 2.1 for muscle and fat [6]. However, it can be seen that the variations in the range from 100 MHz to approximately 1 GHz are relatively small. There are also variations in tissue properties with age because of the variation in water content [9].

However, the penetration depth does change significantly across this frequency range, as shown in Figure 2.2, as derived from (2.1). At 100 MHz, penetration depths are significant, and thus frequencies in this range are used for penetration into the body for communications with medical implants. As the frequency increases, the penetration depth reduces. At 2.45 GHz, the penetration depth is 113 and 21 mm for fat and muscle, respectively. Therefore, in general, penetration at these frequencies is small, and propagation will be around the surface of the body.

2.2 Physical Body Phantoms

A phantom can be defined as a simulated biological body or as a physical model simulating the characteristics of the biological tissues. The aim of such a phantom is to explore the interaction between the human tissue and the electromagnetic fields. For this purpose, phantoms have been used extensively in medical research on the effects of electromagnetic radiation on health, as well as in development of various methods of medical diagnosis and treatment, such as X-ray, magnetic resonance imaging (MRI) scan, and hyperthermia.

With the proliferation of communication devices that will be used close to the human body, and especially with the advent of mobile phones, the phantoms became an essential tool for testing safety of such devices. Various safety guidelines, such as those by the International Commission on Non-Ionizing Radiation Protection (ICNIRP) [10] and the Institute of Electrical and Electronics Engineers (IEEE) [11], specify the acceptable levels of radiation in terms of specific absorption rate (SAR), which can be measured using a number of methods involving phantoms.

Table 2.1 Parameters Needed to Find Body Tissue Dielectric Constant and Conductivity at Any Frequency

	ε_∞	$\Delta\varepsilon_1$	τ_1 (ps)	α_1	$\Delta\varepsilon_2$	τ_2 (ns)	α_2	σ	$\Delta\varepsilon_3$	τ_3 (ns)	α_3	$\Delta\varepsilon_4$	τ_4 (ms)	α_4
Aorta	4.000	40.00	8.842	0.100	50	3.183	0.100	0.250	1.00E+5	159.155	0.200	1.00E+7	1.592	0.000
Bladder	2.500	16.00	8.842	0.100	400	159.155	0.100	0.200	1.00E+5	159.155	0.200	1.00E+7	15.915	0.000
Blood	4.000	56.00	8.377	0.100	5200	132.629	0.100	0.700	0.00E+0	159.155	0.200	0.00E+0	15.915	0.000
Bone (cancellous)	2.500	18.00	13.263	0.220	300	79.577	0.250	0.070	2.00E+4	159.155	0.200	2.00E+7	15.915	0.000
Bone (cortical)	2.500	10.00	13.263	0.200	180	79.577	0.200	0.020	5.00E+3	159.155	0.200	1.00E+5	15.915	0.000
Brain (gray matter)	4.000	45.00	7.958	0.100	400	15.915	0.150	0.020	2.00E+5	106.103	0.220	4.50E+7	5.305	0.000
Breast fat	2.500	3.00	17.680	0.100	15	63.660	0.100	0.010	5.00E+4	454.700	0.100	2.00E+7	13.260	0.000
Cartilage	4.000	38.00	13.263	0.150	2500	144.686	0.150	0.150	1.00E+5	318.310	0.100	4.00E+7	15.915	0.000
Cerebro spinal fluid	4.000	65.00	7.958	0.100	40	1.592	0.000	2.000	0.00E+0	159.155	0.000	0.00E+0	15.915	0.000
Cornea	4.000	48.00	7.958	0.100	4000	159.155	0.050	0.400	1.00E+5	15.915	0.200	4.00E+7	15.915	0.000
Eye tissues (sclera)	4.000	50.00	7.958	0.100	4000	159.155	0.100	0.500	1.00E+5	159.155	0.200	5.00E+6	15.915	0.000
Fat (average infiltrated)	2.500	9.00	7.958	0.200	35	15.915	0.100	0.035	3.30E+4	159.155	0.050	1.00E+7	15.915	0.010
Gall bladder bile	4.000	66.00	7.579	0.050	50	1.592	0.000	1.400	0.00E+0	159.155	0.200	0.00E+0	15.915	0.200
Heart	4.000	50.00	7.958	0.100	1200	159.155	0.050	0.050	4.50E+5	72.343	0.220	2.50E+7	4.547	0.000
Kidney	4.000	47.00	7.958	0.100	3500	198.944	0.220	0.050	2.50E+5	79.577	0.220	3.00E+7	4.547	0.000
Liver	4.000	39.00	8.842	0.100	6000	530.516	0.200	0.020	5.00E+4	22.736	0.200	3.00E+7	15.915	0.050
Lung (inflated)	2.500	18.00	7.958	0.100	500	63.662	0.100	0.030	2.50E+5	159.155	0.200	4.00E+7	7.958	0.000
Muscle	4.000	50.00	7.234	0.100	7000	353.678	0.100	0.200	1.20E+6	318.310	0.100	2.50E+7	2.274	0.000
Skin (dry)	4.000	32.00	7.234	0.000	1100	32.481	0.200	0.000	0.00E+0	159.155	0.200	0.00E+0	15.915	0.200
Skin (wet)	4.000	39.00	7.958	0.100	280	79.577	0.000	0.000	3.00E+4	1.592	0.160	3.00E+4	1.592	0.200
Small intestine	4.000	50.00	7.958	0.100	10000	159.155	0.100	0.500	5.00E+5	159.155	0.200	4.00E+7	15.915	0.000
Stomach	4.000	60.00	7.958	0.100	2000	79.577	0.100	0.500	1.00E+5	159.155	0.200	4.00E+7	15.915	0.000
Testis	4.000	55.00	7.958	0.100	5000	159.155	0.100	0.400	1.00E+5	159.155	0.200	4.00E+7	15.915	0.000
Tongue	4.000	50.00	7.958	0.100	4000	159.155	0.100	0.250	1.00E+5	159.155	0.200	4.00E+7	15.915	0.000

Table 2.2 Electromagnetic Properties of Human Body Tissue at 2.45 GHz

Tissue Name	Conductivity [S/m]	Relative Permittivity	Loss Tangent	Penetration Depth [m]
Aorta	1.467	42.47	0.24837	0.023761
Bladder	0.69816	17.975	0.27927	0.032545
Blood	2.5878	58.181	0.31981	0.015842
Bone, Cancellous	0.82286	18.491	0.31996	0.028087
Bone, Cortical	0.40411	11.352	0.25597	0.044616
Brain, Gray Matter	1.843	48.83	0.27137	0.02031
Breast Fat	0.14067	5.137	0.1969	0.085942
Cartilage	1.7949	38.663	0.3338	0.018638
Cerebro Spinal Fluid	3.5041	66.168	0.38078	0.012537
Cornea	2.3325	51.533	0.32544	0.016548
Eye Sclera	2.0702	52.558	0.28321	0.018773
Fat	0.10672	5.2749	0.14547	0.11455
Gall Bladder Bile	2.8447	68.305	0.29945	0.015592
Heart	2.2968	54.711	0.30185	0.017286
Kidney	2.4694	52.63	0.33736	0.015811
Liver	1.7198	42.952	0.2879	0.020434
Lung, Inflated	0.81828	20.444	0.28779	0.02963
Muscle	1.773	52.668	0.24205	0.021886
Skin, Dry	1.4876	37.952	0.28184	0.022198
Skin, Wet	23.984	20.369	0.84665	0.0010736
Small Intestine	3.2132	54.324	0.42529	0.012438
Stomach	2.2546	62.078	0.26114	0.018707
Testis	2.2084	57.472	0.27628	0.018394
Tongue	1.8396	52.558	0.25167	0.021083

Phantoms are also capable of being a valuable tool in the study of radiowave propagation around and inside the human body. Such studies are required in order to design powerful, robust, wearable low-cost communication devices. The use of phantoms can provide a stable, controllable propagation environment, which cannot be easily realized with human subjects.

The phantoms, in general, can be classified from several different points of view. One classification criterion is the frequency range. Another criterion is the type of tissue that the phantom represents. There are two main types of tissue. The first type is low-water content tissue, such as bones and fats, which feature low permittivity and low loss. The second type is high-water content tissue, such as brain, muscles, and skin, which have higher permittivity and loss. The most common classification is that based on the final state of the phantom after the manufacturing process is completed, which can be either solid (dry), semisolid (gel), solid (wet), or liquid.

2.2.1　Liquid Phantoms

Liquid phantoms are the first and oldest of all the phantom types. It is basically a container filled with a liquid that has the same electrical characteristics as the tissues

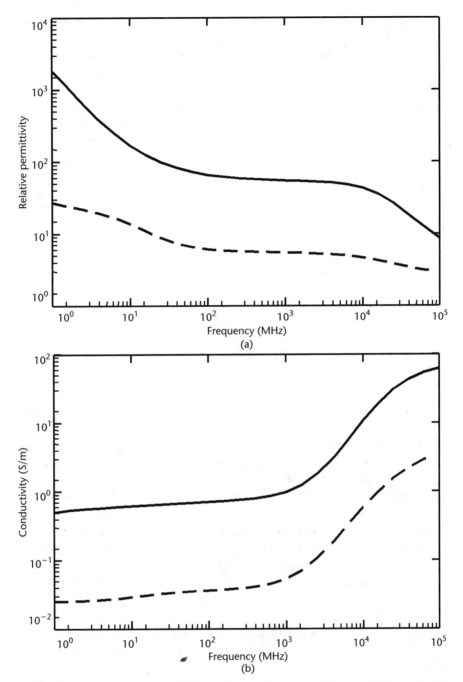

Figure 2.1   Electrical characteristics of tissues [8]: (a) relative permittivity; and (b) conductivity. Solid: muscle; dotted: fat.

in the human body, in the defined range of frequencies. Liquid phantoms are used extensively in SAR studies, using direct measurement of the electric fields inside the phantom with a small probe. The phantoms used in this method have a form of a thin shell, usually made of a fiberglass material with low relative permittivity and conductivity. As specified in [12], the relative permittivity of the phantom shell

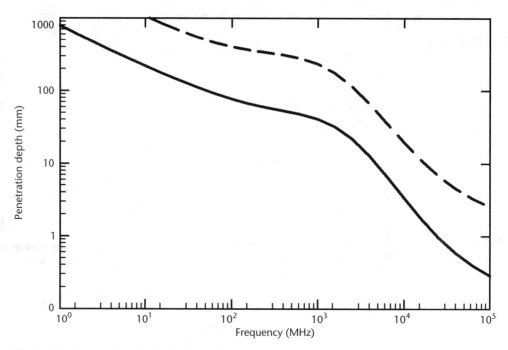

Figure 2.2 Penetration depths for fat and muscle. Solid: muscle; dotted: fat.

should be less than 5, the loss tangent should be less than 0.05, and the thickness of the shell should be 2.0 ± 0.2 mm for the frequencies in the range from 0.8 to 3 GHz. The shell has a shape of a human head or a whole body, depending on the requirements of the application. In some cases, a flat rectangular prism shape is considered sufficient. The shell must have a hole into which the measurement probe is inserted. The phantom is filled with a liquid that has the dielectric properties similar to those of a human body. Most recipes for the liquid contain either sugar at lower frequencies, or, at higher frequencies, diacetin or di-ethylene glycol butyl ether (DGBE) in different proportions to control permittivity of the solution, while salt (NaCl) is used to adjust conductivity [13].

This type of phantom allows the recording of the detailed distribution of the fields inside the phantom. However, it does not allow measurement of SAR close to the surface of the body. In addition, the human body is not represented accurately, since the internal structure is replaced with a homogeneous medium. The latter problem is usually dealt with by choosing the dielectric properties of the phantom liquid such that it leads to the peak spatial-average SAR values that are no less than those produced by the real internal structure of the human body. For the studies of propagation around the body, this is less of an issue, as long as the losses in the body are large, in which case, only a relatively thin external layer of the phantom has a significant effect on propagation. This type of phantom also suffers from such disadvantages as the limited range of frequencies over which the liquid can have the required dielectric properties, the difficulty of handling the container in the test environment, and the electrical characteristics of the container itself. Although this type of phantom has these disadvantages, it still has the important advantage of being easy to fabricate.

2.2.2 Semisolid (Gel) or Solid (Wet) Phantoms

In order to eliminate the outer shell of the phantoms, a coagulant has been used to make solid (wet) phantoms, which are capable of self-shaping. Guy [13] has developed a gel phantom, which is composed of water, sodium chloride, TX-150 (polyamide resin), and polyethylene powder. Here, polyethylene powder and sodium chloride are used to control the relative permittivity and conductivity of the material, respectively. In addition, TX-150 is a gelling agent, which can realize the mixture between water and polyethylene powder. Ito et al. [14] have developed the self-shaping phantom based on Guy's recipe. They have added sodium dehydroacetate and agar, which are used as preservative and coagulant, to Guy's recipe. These types of phantoms are suitable only for simulating high-water content materials (muscle, brain, and so forth) and adjust the electrical characteristic over a wide frequency range. Some references refer to this type as a solid or gel type.

Another solid gel phantom material uses polyacrylamide as its main constituent; this needs special care because of its toxicity. This type of material is capable of simulating both high- and low-water content materials, depending on the liquid solvent used in its fabrication. The material is transparent and can be used in a wide frequency range, up to 5.5 GHz.

All the wet phantom materials degrade over time, due to the loss of water and/or the growth of fungi.

2.2.3 Solid (Dry) Phantoms

If the internal structure of the body has to be preserved in the phantom, or if the SAR on the surface of the body has to be measured, then the use of a solid phantom is a good option. A solid phantom is made of materials that are capable of keeping their shape for a period of time. The SAR measurements on such phantoms are performed by the method of thermography [15]. Before the measurement, the phantom has to be precut along the plane in which the SAR is to be measured. Then, after the phantom is illuminated by a radio source, the precut parts of the phantom are quickly separated, and an infrared sensor measures their surface temperature. The SAR then is derived from the temperature rise. This type of phantom is also best suited for the studies of the propagation around, as well as inside, the body, since it can accurately represent the inhomogeneous structure of the human body.

A number of recipes for solid phantoms have been proposed, including a mixture of ceramic and graphite powder [16], silicone rubber mixed with carbon fiber [17], and conductive plastic containing carbon black [18]. These phantoms have excellent mechanical and dielectric properties that do not degrade over a long time, since they do not contain water. The main disadvantage of ceramic and graphite phantoms results from the fact that this kind of phantom needs special, expensive equipment, along with special procedures in the production of the composition, such as very high temperature (260°C) and high pressure. The other phantoms are easier to fabricate, but it can be difficult to produce large quantities of these materials due to their short curing times.

2.2.4 Examples of Physical Phantoms

In the measurement of the characteristics of antennas close to the human body, head, abdomen, torso, or whole-body phantoms are usually used. In particular, the characteristics of antennas for cellular phones are very dependent on the shape and size of the phantoms [19, 20]. It is therefore necessary to use an optimized phantom to measure the characteristics of antennas. This section describes examples of such phantoms.

Ogawa et al. [19, 20] proposed a multipostural realistic human torso phantom, which is composed of a shell filled with liquid, to evaluate characteristics of antennas. Figure 2.3(a, b) shows two multipostural realistic human torso phantoms. The size of the phantom is based on the average values of Japanese males in their twenties. The phantom can be applied to the various situations, such as talking [Figure 2.3(a)] or viewing [Figure 2.3(b)]. The measured radiation characteristics of the antenna close to the phantom are almost equal to those of real humans.

Hall et al. [21] used a whole-body human phantom, based on the 95th percentile of the anthropomorphic data collected by U.S. Army, to evaluate SAR inside the body. The phantom is sectional, with removable arms and legs, and has been built in two configurations, seated and standing. The seated version is intended specifically for the measurement inside vehicles.

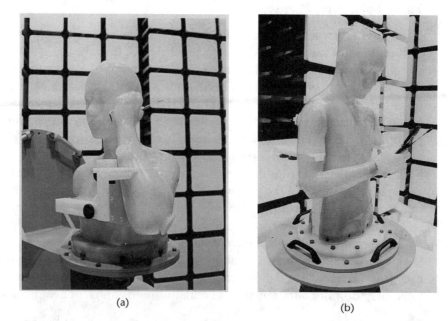

(a)   (b)

Figure 2.3 Multipostural realistic human torso phantom: (a) talk position and (b) viewer position. (Courtesy of Dr. K. Ogawa, Matsushita Electric Industrial Co., Ltd., Japan.)

Figure 2.4 shows a human torso phantom of a Japanese male, which is composed of an agar-based phantom [14, 22, 23], to evaluate electromagnetic (EM) dosimetry, as well as the characteristics of antennas. This phantom, which is called the TX-151 phantom, can simulate the dielectric constants of high-water content tissues from 800 to 2,400 MHz [22], and from 3.8 to 5.8 GHz [23]. Other features of the phantom are summarized as follows:

- Easy adjustment of dielectric properties;
- Easy manufacturing to arbitrary shape and multiple layers [22];
- No necessity to use a container or shell;
- Cheap and easily obtained ingredients;
- Life expectancy of about one month with wrapping.

Table 2.3 shows an example of the composition of a head-equivalent solid phantom [23]. Here, the dielectric properties are adjusted by the amount of polyethylene powder (relative permittivity) and the amount of sodium chloride (conductivity), respectively. The sodium dehydroacetate is a preservative. The agar is a coagulant. Figure 2.5 shows an example of the dielectric constant of the phantom against frequencies over the range from 3 to 6 GHz [23]. The dielectric constant of the phantom agrees with those of the target within 5% over the range from 3.8 to 5.8 GHz.

Okano et al. [24] proposed the high-preservative phantom with a lifespan greater than six months, which is called a glycerin phantom. This phantom can simulate tissues with both high-water and low-water content. However, the useful frequency range of this phantom is narrower than that of the TX-151 phantom. Kobayashi et al. [15] developed a steady solid phantom, which is called the dry

Figure 2.4   Realistic human torso phantom (agar-based solid phantom).

Table 2.3 Example of Composition of Head-Equivalent Phantom [23]

Ingredients	Amount [g]
Deionized water	3,375
Sodium chloride	60.8
Polyethylene powder	675
TX-151	45.6
Sodium dehydroacetate	2.0
Agar	104.6

A batch is approximately 4,000 cm³.

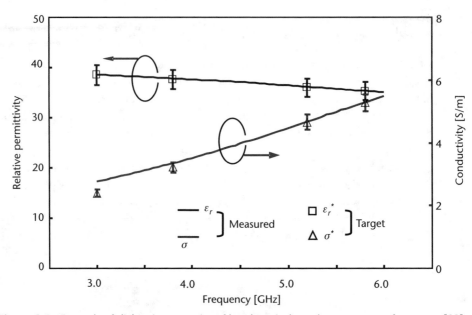

Figure 2.5 Example of dielectric properties of head-equivalent phantom versus frequency [23].

phantom. The dielectric properties of this type are not dependent on time. However, the manufacturing of this phantom is fairly difficult in most conventionally equipped laboratories. Figure 2.6 shows an example of a human-head dry phantom.

2.3 Numerical Phantoms

Many numerical phantoms for theoretical analyses and computational simulations have been reported. In theoretical analyses, simple-shaped phantoms are generally used. These types are called theoretical phantoms. However, in order to calculate the characteristics of antennas close to the human body, it is necessary to use a more realistic numerical phantom, which is composed of many voxels. Therefore, such phantoms are called voxel phantoms. This section shows examples of both theoretical and voxel phantoms.

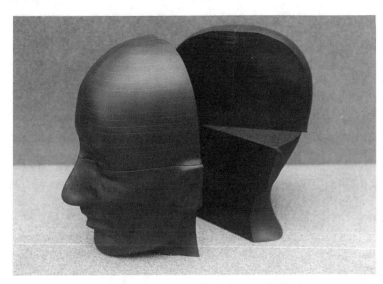

Figure 2.6   Example of dry phantom. (Courtesy of Professor T. Nojima, Hokkaido University, Japan.)

2.3.1   Theoretical Phantoms

Homogeneous or layered flat phantoms [25] are the simplest theoretical phantoms, and are used to evaluate EM dosimetry, in which energy is radiated from simple sources, such as a plane wave, halfwave dipole, or small dipole antenna. Spherical models are mainly used for EM dosimetry inside the human head [26–29]. The spherical models are also used for dosimetry in the eyes. Cylindrical phantoms [30, 31] are used for whole-body models.

These models can also be used for the confirmation of the validity of results using numerical simulation tools, such as the finite difference time domain (FDTD) method [32], and the method of moments (MoM) [33]. Other canonical models, such as a 200^3-mm^3 cube and a 200-mm diameter sphere, have been used [34].

2.3.2   Voxel Phantoms

Recently, progress in medical imaging technologies, such as MRI and X-ray computed tomography (CT), and large increases in computer power, have encouraged the development of precision-head and whole-body voxel models.

Human-head models are usually used for the evaluation of EM dosimetry, as well as characteristics of antennas, when cellular phones are placed close to the head [35, 36]. In addition, the evaluation of SAR inside head models of children has been also proposed [37–39]. However, the internal structure of the child-head models based on the adult models can be significantly different from the real child [40].

Whole-body human voxel models have been proposed. Dimbylow [41] developed an anatomically realistic voxel model of an entire body to simulate the reference man (176 cm, 73 kg) described in ICRP 23 [42]. This model, which is based on magnetic resonance images (MRIs) of an adult male, was named NORMAN (normalized man), and was segmented into 37 different types of tissues. The resolution of NORMAN is $2.04 \times 2.04 \times 1.95$ mm^3.

Mason et al. [43] proposed a very high spatial resolution whole-body human voxel model (voxel size = 1 mm³), classifying over 40 different types of tissues, based on photographic male data from the visible human project (VHP) dataset at the U.S. National Library of Medicine [44]. Figure 2.7 shows the whole-body human voxel model [45]. The original data of the VHP was obtained from a 38-year-old male cadaver, 186 cm in height and 90 kg in weight, while the voxel model weighs 105 kg, due to the process of modeling. This model is widely used for numerical simulation throughout the world.

Nagaoka et al. [46] developed high-resolution whole-body Japanese human voxel models (voxel size = 2³ mm³), classifying 51 different types of tissues, based on images of males and females that have approximately the same body size as the average Japanese human aged 18 to 30 years. Figure 2.8(a, b) shows whole-body Japanese human voxel models. Here, the 22-year-old male was 172.8 cm tall and weighed 65.0 kg; the 22-year-old female was 160.0 cm tall and weighed 53.0 kg. In addition, the adult female model is the first of its kind in the world, and both are the first Asian voxel models that enable numerical evaluation of electromagnetic dosimetry at higher frequencies, up to 3 GHz.

Kainz et al. [47] developed the abdomen model of a pregnant woman. Here, the structure of the human fetus is based on MR images. In the near future, whole-body human-child and pregnant women models will be developed by researchers.

2.4 Numerical Modeling Techniques for Body-Centric Wireless Communications

2.4.1 Introduction of Numerical Techniques

Past attempts at numerical modeling of human bodies were mainly in the area of calculating the SAR for mobile phones, and the shadowing effects for indoor wireless networks. However, numerical modeling for on-body antennas and radio propaga-

Figure 2.7   High spatial resolution whole-body human voxel model. (*From:* [45]. © 2000 The Health Physics Society. Reprinted with permission.)

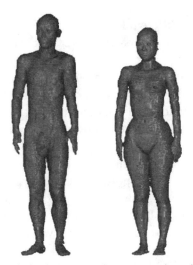

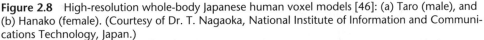

Figure 2.8   High-resolution whole-body Japanese human voxel models [46]: (a) Taro (male), and (b) Hanako (female). (Courtesy of Dr. T. Nagaoka, National Institute of Information and Communications Technology, Japan.)

tion has distinct features, particularly at microwave frequencies. In this case, the human body is treated as a transmission medium that guides the surface wave, and as a reflector for space waves. Usually, when more than two communication nodes are placed on the body, or in close proximity to the body, changes in the polarizations and orientations of the antennas and the body postures significantly affect the radio propagation paths and cause large variations in path gain. In addition, in on-body communication systems, the dimensions of the body are large relative to the wavelength. This allows some semianalytical approaches, and the approximation of the body as a homogenous lossy dielectric, to be used. The following sections briefly outline several frequently used numerical modeling techniques for analyzing electromagnetic problems, especially in the area of body-centric wireless communications.

2.4.1.1   The Uniform Geometrical Theory of Diffraction

The Uniform Geometrical Theory of Diffraction (UTD), and its earlier version, the Geometrical Theory of Diffraction (GTD), are based on geometric optics (GO) and diffraction theory [48]. It assumes that all waves are local plane waves, so that ray tracing (RT) can be used. GTD and UTD require that objects under study are electrically large, as is the case for the human body in Bluetooth and WLAN frequency bands.

Reflection and transmission (or penetration) is the basic phenomenon when the interactions between propagation waves and the surrounding environment occur. Both coefficients can be calculated by classic formulations, depending on the polarization of incidence. Wedge and corner diffractions are unique in methods like UTD. Diffraction is a form of scattering by objects whose size is of the same order of magnitude as the wavelength. Inclusion of various diffractions in UTD modeling is essential, otherwise abrupt changes in field strength may occur if the receiver moves from line-of-sight (LOS) to nonline-of-sight (NLOS) situations. There is also a kind of diffraction that is different from GO, called creeping waves. A ray incident tangential to a curved surface sets off a creeping ray that moves along a geodesic of the surface.

In [49], a hybrid uniform asymptotic theory of diffraction/finite difference time domain (UTD/FDTD) method was proposed to analyze human exposure in realistic urban environments for the Global System for Mobile communications (GSM) and the Universal Mobile Telecommunications System (UMTS) frequency bands. In [50], a practical deterministic propagation prediction model is introduced to investigate the human body-scattering effects in the indoor channel using UTD. In the model, the human body is approximated by a perfect conducting circular cylinder, and then combined with the ray-tracing technique to deal with particular indoor propagation scenarios. In later sections, an extension of UTD to on-body radio propagation modeling at 2.4 GHz and UWB frequencies will be discussed.

2.4.1.2   Ray Tracing Techniques

Ray tracing (RT) techniques are popular for the site-specific prediction of radio channel characteristics of wireless communication systems, such as cellular phone services, personal communication systems (PCS), and WLAN. RT provides time delay and angle of arrival (AoA) information for multipath reception conditions. It can be used to accurately determine received signal strength and power delay profiles (PDP) in situations where the wavelength is small relative to propagation environment feature size. It has been widely used to predict indoor radio propagation. There are two types of RT models. One is the image method, and the other is the method of shooting and bouncing rays (SBR) [51]. The image method is well suited to analysis of radio propagation associated with geometries that have low complexity and a small numbers of reflections. The SBR method launches a bundle of rays that may or may not reach the receiver. It requires numerous ray-object intersection tests and extensive data arrays for RT. In the SBR method, both refraction and diffraction can be considered. Due to the complexity of the situations in the on-body environment, only the SBR method is applied, and this is introduced in later sections.

Villanese et al. [52] presented a novel model for indoor wireless communication, based on a dual image and ray-shooting approach. This model, which is capable of improved site-specific indoor propagation prediction, considers multiple human bodies moving within the environment in a modern office, at 2.45 GHz. In [53], a technique was proposed combining the RT method and the FDTD method for the SAR calculation in human bodies that are nonuniformly exposed to the EM fields from indoor cellular base-station antennas.

2.4.1.3   Method of Moments

The MoM is a technique for solving complex integral equations by reducing them to a system of simpler linear equations. It employs the weighted residuals technique introduced by Harrington [33]. The fundamental operation in this technique is to set up a set of trial solution functions with one or more variable parameters. The residuals are a measure of the difference between the trial solution and the true solution. The variable parameters are determined in a manner that guarantees a best fit of the trial functions based on a minimization of the residuals. The MoM, which can be used in both time and frequency domains, is most suitable in the analysis of thin-wire structures.

However, the MoM technique is not effective when applied to arbitrary configurations with complex geometries or inhomogeneous dielectrics. In on-body applications, Chen et al. [54–56] presented an extensive computer simulation of the influence of the human body on a circular-loop-wire antenna to simulate the pager antenna. The coupled integral equations (CIE) approach and the MoM are employed for numerical simulation of this antenna-body-coupling problem. The magnetic frill source is used to model the antenna-feeding structure. A realistically shaped, full-scale human-body model (1.7m) is constructed. A small loop antenna (loop radius $b = 1.7$ cm and wire radius $a = 0.072$ cm) of x, y, and z orientation, in free space or proximate to the human body at the top pocket (chest position) or belt level (waist position), is considered at 152, 280, and 400 MHz for radio paging communications. A hybrid finite element method (FEM)/MoM technique was employed in [57] for SAR calculations in a human phantom in the near field of a typical GSM base-station antenna. The MoM is used to model the metallic surfaces and wires of the base-station antenna, and the FEM is used to model the heterogeneous human phantom. The advantages of each of these frequency domain techniques are exploited, leading to a highly efficient and robust numerical method for addressing this type of bioelectromagnetic problem.

2.4.1.4   Finite Element Method

The Finite Element Method (FEM) is usually formulated on a variational expression, which has a rigorous mathematical formulation in terms of convergence [58]. The basic idea of the FEM is to divide the electromagnetic structures into a number of elements in various shapes, such as rectangular or triangular. The field within each element is expanded in terms of a set of basis polynomials, weighted by the field values at the nodes of the element. Then the expanded field is substituted into the functional for the Maxwell's equations, and the variation of the functional is set to zero. A matrix eigenvalue equation is derived, with the field values at the element nodes as the unknowns, and the solution allows the eigenvalues and eigenvectors to be determined.

FEM is well suited to modeling electromagnetic structures with curved boundaries, which can be represented using triangular and isoparametric elements. A major disadvantage of FEM is that the matrix solution is quite time-consuming, even if the sparse matrix technique is employed. As a result, the computational complexity turns out to be the highest among all the available approaches.

FEM has been applied mostly at low megahertz frequencies for medical applications, due to computer memory constraints. In [59], an FEM-based model was adopted for calculating eddy current effects in the human body using magnetic resonance techniques in modern medical diagnosis. The frequency of the linearly polarized RF field was 64 MHz. The body components considered in the models included skin, fat, bone, muscle, marrow, spinal cord, liver, kidney, bladder, bowels, and connective tissue. It was concluded that a coarser discretization of the innermost organs was sufficient.

2.4.1.5 Finite-Difference Time-Domain (FDTD) Method

The FDTD method is one of the best-known numerical methods for solving boundary-value problems, and has been applied to the electromagnetic problems for some years. Like FEM, it demands the division of the EM structures into a set of small cells, and therefore is suitable for modeling inhomogeneous media and complicated boundaries.

Yee proposed the FDTD method in 1966 [32]. Extensions and enhancements to the method are continuously announced, which can be seen from the number of published papers counted by Schneider [60]. Since 1991, there have been a number of research papers on the application of the FDTD method in calculating SAR for mobile applications [61]. Modifications on the conventional FDTD scheme have also been made in order to achieve a memory-efficient human model. Lazzi et al. [62] proposed a method that uses prestored FDTD-computed impulse responses (Green's functions) of the human body model, and integrates them with the complex incident electromagnetic field distribution that can be measured on site. The application of this method to the dosimetry of various cellular telephone base station antennas was presented, showing its versatility and ease of use. Early efforts used crude models of the human head and body. In [63], a spherical head approximation was used near a simple FDTD mesh model of a portable telephone. In [64], a slightly more realistic human head model was used, again with a nearby portable telephone. An approximate FDTD mesh for an entire human body was shown in [65]. Such models are only useful for predicting antenna patterns, and for rough estimates of the electromagnetic fields inside a human head or body. In order to provide more accurate estimates of the internal electromagnetic fields, more realistic FDTD meshes have been obtained from MRI scan data [66]. An alternative approach for obtaining the human body data is the Visible Human Project. This project involved slicing a frozen cadaver at 1-mm intervals, photographing the slices, and digitizing the data. The result is 1,871 digital files, one for each cross section, at a resolution of 2,048 × 1,216 pixels, with 24-bit color. The total amount of data is approximately 15 GB. A convenient source for information on this data is available at http://www.nlm.nih.gov, the home page of the National Library of Medicine [44, 67].

One of the major reasons for the rapid growth of the FDTD method is that it requires only $O(N)$ computer complexity, while MoM and FEM need $O(N^2)$ computational storage, where N is the number of the unknowns. Wideband results, useful for UWB body-centric systems, can be obtained by performing a fast Fourier transform on the time-domain response, which is available with only one simulation.

As with other numerical techniques, the FDTD method has a few weaknesses. First, it requires the entire computational domain to be meshed, and these cells must be small compared to the smallest wavelength, and smaller than the smallest feature in the model. The former results in the large computational burden and the latter makes the modeling of long and thin structures very difficult. Second, because the many elements on electromagnetic structures are curved, the edges will not be smooth, but be staircased. Where configurations have sharp or acute-angled edges, an adequately stair-cased approximation may require a very small grid size and, consequently, a small time step. In this respect, FDTD is not as flexible as the FEM method.

2.4.2 On-Body Radio Channel Modeling

2.4.2.1 A Brief Introduction of Modeling Techniques

Accurate prediction of radio propagation behavior is crucial to body-centric wireless system design. It is important to optimize wearable node locations to ensure satisfactory system coverage. Site measurements have the advantage of accounting for all parameters without presumptions. However, they are expensive and time-consuming. Therefore, it is necessary to develop effective body-centric channel models. Many statistical and site-specific models for narrowband indoor/outdoor propagation analysis have been established using an RT technique and the FDTD method. Sarkar et al. [68] presented a survey of various propagation models for mobile communications. Wang et al. [69] introduced a hybrid technique based on the combination of RT and FDTD methods, but only applicable for narrowband systems. In 2004, Attiya et al. [70] proposed a simulation model for UWB systems using RT. However, the accuracy of this model may decrease when small scatterers are considered. When applying existing narrowband channel models to UWB systems, a number of issues have to be taken into account. For example, for a UWB system, human tissues and materials in the local environment are likely dispersive across a wide frequency band; for example, the UWB, which ranges from 3.1 to 10.6 GHz. To accurately include material information in the simulations, one solution is to apply the dispersive FDTD method [71]; another solution is using a newly proposed subband FDTD method. In this section, several modified modeling techniques are introduced in detail, namely, the UTD, RT, conformal FDTD, and dispersive FDTD.

2.4.2.2 Conformal FDTD on On-Body Radio Propagation Modeling

The locally distorted, nonorthogonal FDTD (LN-FDTD) method [72] was developed so that only those cells close to the curved boundary are distorted and processed with the nonorthogonal FDTD (NFDTD) algorithm; others remain as Cartesian cells. As an example, the iteration of a transverse electric (TE) polarization problem is given. Suppose at a cell node (i, k) there is a Cartesian cell. Then, the conventional Yee algorithm for this cell can be implemented:

$$E_x^{n+1}(i,k) = E_x^n(i,k) - \frac{dt}{\varepsilon \cdot dz} \cdot \left[H_y^{n+1/2}(i,k) - H_y^{n+1/2}(i,k-1) \right] \tag{2.2}$$

$$E_z^{n+1}(i,k) = E_z^n(i,k) + \frac{dt}{\varepsilon \cdot dx} \cdot \left[H_y^{n+1/2}(i,k) - H_y^{n+1/2}(i-1,k) \right] \tag{2.3}$$

$$H_y^{n+1/2}(i,k) = H_y^{n-1/2}(i,k) + \frac{dt}{\mu \cdot dx} \cdot \left[E_z^n(i+1,k) - E_z^n(i,k) \right]$$
$$- \frac{dt}{\mu \cdot dz} \cdot \left[E_x^n(i,k+1) - E_x^n(i,k) \right] \tag{2.4}$$

where dx and dz are the grid increments in the x and z direction, respectively. If the node (l, m) pertains to a distorted cell, then Holland's equations [73] can be used to find the electromagnetic properties of that cell:

$$E^x(l,m)^{n+1} = E^x(l,m)^n$$

$$- \frac{dt}{\varepsilon} \cdot \left[\frac{H_y(l,m)^{n+1/2}}{\sqrt{g^{22}(l,m)}} - \frac{H_y(l,m-1)^{n+1/2}}{\sqrt{g^{22}(l,m-1)}} \right] \sqrt{\frac{g_{11}(l,m)}{g(l,m)}} \qquad (2.5)$$

$$E^z(l,m)^{n+1} = E^z(l,m)^n + \frac{dt}{\varepsilon} \cdot \left[\frac{H_y(l,m)^{n+1/2}}{\sqrt{g^{22}(l,m)}} - \frac{H_y(l-1,m)^{n+1/2}}{\sqrt{g^{22}(l-1,m)}} \right] \sqrt{\frac{g_{33}(l,m)}{g(l,m)}} \qquad (2.6)$$

$$H^y(l,m)^{n+1/2} = H^y(l,m)^{n-1/2} + \frac{dt}{\mu} \cdot \sqrt{\frac{g_{22}(l,m)}{g(l,m)}}$$

$$\cdot \left[\frac{E_z(l+1,m)^n}{\sqrt{g^{33}(l+1,m)}} - \frac{E_z(l,m)^n}{\sqrt{g^{33}(l,m)}} - \frac{E_x(l,m+1)^n}{\sqrt{g^{11}(l,m+1)}} + \frac{E_x(l,m)^n}{\sqrt{g^{11}(l,m)}} \right] \qquad (2.7)$$

On the left-hand side of the above equations, the components being updated are the contravariant E and H fields, whereas on the right-hand side, there are the covariant components. To complete the iteration for these components, an interpolation scheme is required, as follows

$$E_x(l,m) = G_{11}(l,m) \cdot E^x(l,m) + \frac{G_{13}(l,m)}{4}$$

$$\cdot \left[\begin{array}{c} E^z(l+1/2, m-1/2) + E^z(l-1/2, m-1/2) \\ +E^z(l+1/2, m+1/2) + E^z(l-1/2, m+1/2) \end{array} \right] \qquad (2.8)$$

and similarly for E_z. For the third component, we have $H_y = H^y$. In the above, G is a matrix related to the metric tensor by

$$G_{ij} = \sqrt{\frac{g_{ii}}{g_{jj}} \cdot g_{ij}} \qquad (2.9)$$

At the interface between the Cartesian and the distorted cells, the components match each other automatically by letting

$$E,H_{\text{Cartesian}} = E,H_{\text{Covariant}} = E,H_{\text{Contravariant}} \qquad (2.10)$$

Therefore, the interpolation at the interface is very simple and does not need any extra computer resources. A complete discussion about the LD-NFDTD method and the calculation of the metric tensor can be found in [74].

The Queen Mary, University of London conformal FDTD modeling software was used to evaluate on-body propagation channels. The human body grid, shown in Figure 2.9, was given the following tissue parameters at 2.4 GHz. The relative permittivity and conductivity of dry skin were 38.1 and 1.441 S/m, respectively; those for muscle, which is used over the whole body, were 52.7 and 1.705 S/m, respectively; and those for lungs, which have a large volume in the upper body, were 34.5 and 1.219 S/m, respectively. Bones and other organs were not considered

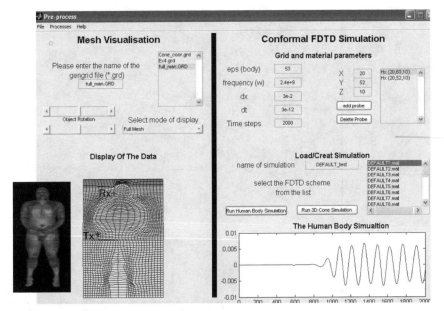

Figure 2.9   A GUI for an in-house conformal FDTD program, showing a human body, grid, tissue settings, and signal responses. The outline shape is similar to the one from invisible man project (inset picture). Tissue parameters are: dry skin, $\varepsilon_r = 38.1$, $\sigma = 1.441$ S/m; muscle, $\varepsilon_r = 52.7$, $\sigma = 1.705$ S/m; and lungs, $\varepsilon_r = 34.5$, $\sigma = 1.219$ S/m.

because of their relatively small volumes. The path loss variation against the distance is shown in Figure 2.10, compared to measurement results for the monopole-loop case for the belt-to-chest case when the body is standing straight up [see Figure 3.13(a)]. Good agreement is seen. The electric field distribution around and on the surface of the human body at different time steps, using this simulation method, can be found in Figure 3.20. The energy is found to travel along the human body surface. Attenuation and shadowing are apparent in later time steps, with the field being stronger at the source side of the body [75].

2.4.2.3   Modified FDTD Techniques for UWB On-Body Channel Modeling

Dispersive FDTD Method   One of the most significant developments in the FDTD method is its capability of modeling frequency-dependent materials. The existing dispersive FDTD methods can be roughly categorized into three types: the differential equation-based method, the Z-transform method, and the discrete convolution method. The choice of method depends on the frequency dependence of material properties. In our studies, the Debye material is considered. The complex frequency-dependent permittivity can be modeled using [76]

$$\varepsilon^*(\omega) = \varepsilon_\infty + \frac{\varepsilon_s - \varepsilon_\infty}{1 + j\omega\tau_0} + \frac{\sigma}{j\omega\varepsilon_0} \tag{2.11}$$

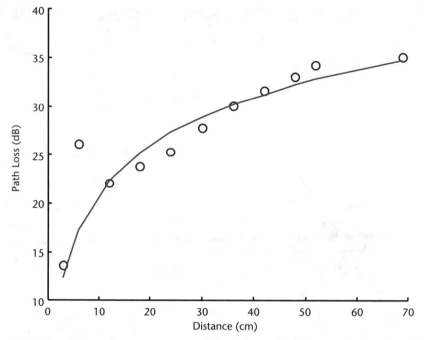

Figure 2.10 Comparison of measurements and simulations of belt monopole-chest loop link when body standing straight up. [Solid line: measurement; dots: conformal FDTD simulation; frequency = 2.45 GHz. See Figure 3.13(a) for details.]

where ε_∞ is the infinite frequency, ε_s is the static permittivity, τ_0 is the relaxation time, and σ is the conductivity. Since the electric flux density $\vec{D}(\omega)$ is related to the electric field $\vec{E}(\omega)$ by

$$\vec{D}(\omega) = \varepsilon^*(\omega) \cdot \vec{E}(\omega) \tag{2.12}$$

one can write

$$j\omega\varepsilon_0\left(1 + j\omega\tau_0\right) \cdot \vec{D}(\omega) = \left[j\omega\varepsilon_0\varepsilon_\infty\left(1 + j\omega\tau_0\right) + \sigma\left(1 + j\omega\tau_0\right) + j\omega\varepsilon_0\left(\varepsilon_s - \varepsilon_\infty\right) \right] \cdot \vec{E}(\omega) \tag{2.13}$$

According to the differentiation property of the Fourier transform, (2.13) can be written in differential form:

$$\varepsilon_0\tau_0\frac{\partial^2 \vec{D}}{\partial t^2} + \varepsilon_0\frac{\partial \vec{D}}{\partial t} = \varepsilon_\infty\varepsilon_0\tau_0\frac{\partial^2 \vec{E}}{\partial t^2} + \left[\varepsilon_\infty\varepsilon_0 + \sigma\tau_0 + \left(\varepsilon_s - \varepsilon_\infty\right)\varepsilon_0\right]\frac{\partial \vec{E}}{\partial t} + \sigma\vec{E} \tag{2.14}$$

In central difference form, it is

$$\frac{\varepsilon_\infty\varepsilon_0\tau_0}{dt^2}\left(E_z^{n+1} - 2E_z^n + E_z^{n-1}\right) + \frac{\varepsilon_\infty\varepsilon_0 + \sigma\tau_0 + \left(\varepsilon_s - \varepsilon_\infty\right)\varepsilon_0}{dt}\left(E_z^{n+1} - E_z^n\right)$$
$$+ \frac{\sigma}{2}\left(E_z^{n+1} + E_z^n\right) = \frac{\varepsilon_0\tau_0}{dt^2}\left(D_z^{n+1} - 2D_z^n + D_z^{n-1}\right) + \frac{\varepsilon_0}{dt}\left(D_z^{n+1} - D_z^n\right) \tag{2.15}$$

Equation (2.15) can be incorporated into the conventional FDTD code.

A Subband FDTD Method   In UWB radio channels, the inherent material dispersion represents the change of permittivity and conductivity with frequency. Such dispersion cannot be generically modeled using existing dispersive FDTD methods based on Debye/Lorentz relations, and thus a subband FDTD is needed. To apply the subband FDTD method, one can follow these steps [71]:

1. Divide the whole frequency band into several subbands, each of which is narrow enough to assume the same frequency characteristics.
2. Use the conventional FDTD method to obtain the time domain delay profiles for each subband.
3. Fourier transform the subband delay profiles into the frequency domain, extract the "accurate" part, and combine them to give new frequency responses.
4. Transform the frequency responses back into the time domain to have a delay profile that is valid over the entire bandwidth.

In the example below, the frequency band of 3 to 9 GHz is divided into 12 subbands, each with a 500-MHz bandwidth. The choice of the number of subbands depends on the accuracy required in modeling the dispersive material properties. For instance, the relative dielectric constant of human muscle ranges from 52.058 at 3 GHz, to 44.126 at 9 GHz, as shown in Figure 2.2 [8]. Twelve subbands are used to match the frequency dispersion curve, by assuming that the dielectric constant within each subband to be constant (obtained at the center frequency of each subband). The overall error from such a curve fitting is less than 1%. Figure 2.11 shows the frequency-dependent dielectric constant and conductivity of human muscle from measurement, and their stair-casing approximation used in the proposed subband FDTD model.

For Step 3, the combination of received signals in the frequency domain can be obtained using:

$$F_r(\omega) = \sum_{i=1}^{N} F_{r,i}(\omega) \cdot A_i(\omega) \tag{2.16}$$

where $F_{r,i}(\omega)$ is the received frequency domain signal at the ith subband, $A_i(\omega)$ is a rectangle window function associated with the bandwidth of the ith subband, and N is the total number of subbands. Finally, the combined frequency domain signal is inverse Fourier transformed into the time domain to obtain the time delay profile.

Figure 2.12 shows the path loss around an elliptical trunk, or midrift, using the subband FDTD model, and UTD/RT model and measurement. The trunk is approximately 80 cm in circumference. Good agreement is achieved when the distance between the transmitter and receiver is small. However, when the distance approaches 40 cm, ripples are observed from the UTD/RT method and measurement, which are caused by the addition or cancellation of two creeping rays, traveling in opposite directions around the trunk. The subband FDTD model fails to accurately predict such phenomena, due to the staircase approximation of the curved surfaces. However, this problem can be alleviated by using the conformal

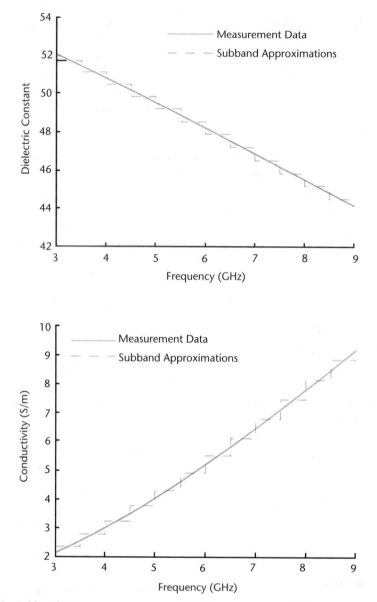

Figure 2.11   Subband approximations for the UTD/RT and subband FDTD models to measure permittivity and conductivity of human muscle.

FDTD method [33]. Figure 2.12 also shows that for modeling simple on-body communication cases, such as with both transmitter and receiver on the trunk, UTD proves to be very efficient and gives accurate results.

Figures 2.13 and 2.14 show the comparison of path loss for different on-body channels using a wideband monopole and printed bowtie antenna, respectively. The calculated path loss exponents are also shown in both figures. The high exponent values are due to the nonreflecting environment (free space). For the monopole, both the subband FDTD and UTD/RT show good agreement with measurements, since the antenna radiation characteristics are relatively stable across the bandwidth. Thus, it is clear that the approximations of using a point source in the subband

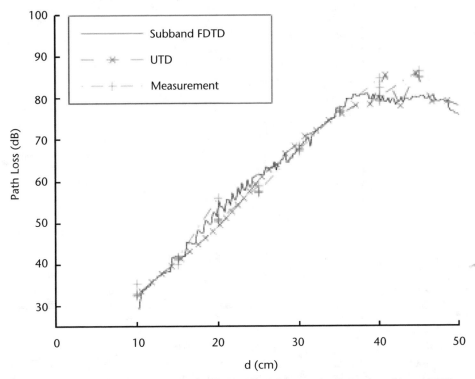

Figure 2.12 Path loss around the human trunk at UWB frequencies, from the subband FDTD model, and the UTD model and measurement.

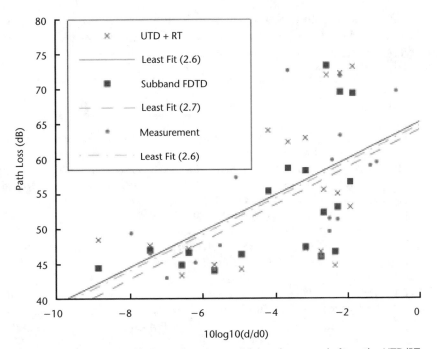

Figure 2.13 Path loss for on-body channels, using a wideband monopole from the UTD/RT model, and subband FDTD model and measurement. Least-square fitted line and path loss exponent values for each model also are shown.

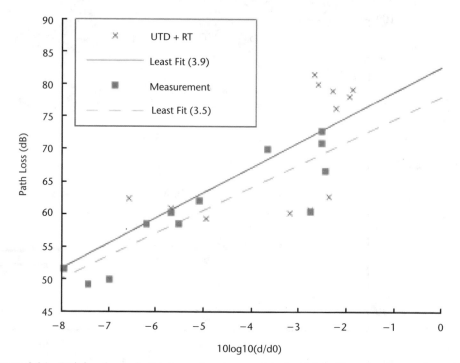

Figure 2.14 Path loss for on-body channels, using a printed wideband bowtie antenna from the UTD/RT model and measurement. Least-square fitted line and path loss exponent values for each model also are shown.

FDTD method and the use of the radiation pattern at 6 GHz for the whole bandwidth in the UTD/RT method are reasonable. For the printed antenna, the larger difference is caused by antenna radiation pattern instability.

References

[1]   Durney, C. H., H. Massoudi, and M. F. Iskander, *Radiofrequency Radiation Dosimetry Handbook*, Brooks Air Force Base-USAFSAM-TR-85-73, 1986.

[2]   Geddes, L. A., and L. E. Barker, "The Specific Resistance of Biological Material—A Compendium of Data for the Biomedical Engineer and Physiologist," *Medical and Biological Engineering*, Vol. 5, 1967, pp. 271–293.

[3]   Stuchly, M. A., and S. S. Stuchly, "Dielectric Properties of Biological Substances—Tabulated," *J. of Microwave Power*, Vol. 15, No. 1, 1980, pp. 19–26.

[4]   Foster, K. R., and H. P. Schwan, "Dielectric Properties of Tissues and Biological Materials: A Critical Review," *Critical Reviews in Biomedical Engineering*, Vol. 17, No, 1, 1989, pp. 25–104.

[5]   Duck, F. A., *Physical Properties of Tissue: A Comprehensive Reference Book*, New York: Academic Press, Harcourt Brace Jovanovich, 1990.

[6]   Gabriel, C., *Compilation of the Dielectric Properties of Body Tissues at RF and Microwave Frequencies*, Brooks Air Force Technical Report, AL/OE-TR-1996-0037, 1996.

[7]   Gabriel, C., T. Y. A. Chan, and E. H. Grant, "Admittance Models for Open-Ended Coaxial Probes and Their Place in Dielectric Spectroscopy," *Physics in Medicine and Biology*, Vol. 39, No. 12, 1994, pp. 2183–2200.

[8]   "An Internet Resource for the Calculation of the Dielectric Properties of Body Tissues," Institute for Applied Physics, Italian National Research Council, http://niremf.ifac.cnr.it/tissprop/.

[9]   Peyman, A., A. A. Rezazadeh, and C. Gabriel, "Changes in the Dielectric Properties of Rat Tissue as a Function of Age at Microwave Frequencies," *Physics in Medicine and Biology*, Vol. 46, 2001, pp. 1617–1629.

[10] ICNIRP, "Guidelines for Limiting Exposure to Time-Varying Electric, Magnetic, and Electromagnetic Fields (Up to 300 GHz)," *Health Physics*, Vol. 74, No. 4, 1998, pp. 494–522.

[11] IEEE, "IEEE Recommended Practice for Determining the Peak Spatial-Average Specific Absorption Rate (SAR) in the Human Head from Wireless Communications Devices: Measurement Techniques," IEEE Std. 1528-2003, 2003.

[12] Kanda, M. Y., et al., "Formulation and Characterization of Tissue Equivalent Liquids Used for RF Densitometry and Dosimetry Measurements," *IEEE Trans. on Microwave Theory and Techniques*, Vol. 52, No. 8, 2004, pp. 2046–2056.

[13] Guy, A. W., "Analyses of Electromagnetic Fields Induced in Biological Tissues by Thermographic Studies on Equivalent Phantom Models," *IEEE Trans. on Microwave Theory and Techniques*, Vol. 19, No. 2, 1968, pp. 205–214.

[14] Ito, K., et al., "Development and Characteristics of a Biological Tissue-Equivalent Phantom for Microwaves," *Electron. Commun. Jpn. Pt. I-Commun.*, Vol. 84, No. 4, 2001, pp. 67–77.

[15] Kobayashi, T., et al., "Dry Phantom Composed of Ceramics and Its Application to SAR Estimation," *IEEE Trans. on Microwave Theory and Techniques*, Vol. 41, No. 1, January 1993, pp. 136–140.

[16] Tamura, H., et al., "A Dry Phantom Material Composed of Ceramic and Graphite Powder," *IEEE Trans. on Electromagnetic Compatibility*, Vol. 39, No. 2, May 1997, pp. 132–137.

[17] Nikawa, Y., M. Chino, and K. Kikuchi, "Soft and Dry Phantom Modeling Material Using Silicone Rubber with Carbon Fiber," *IEEE Trans. on Microwave Theory and Techniques*, Vol. 44, No. 10, Part 2, 1996, pp. 1949–1952.

[18] Chang, J. T., et al., "A Conductive Plastic for Simulating Biological Tissue at Microwave Frequencies," *IEEE Trans. on Electromagnetic Compatibility*, Vol. 42, No. 1, 2000, pp. 76–81.

[19] Ogawa, K., et al., "A High-Precision Real Human Phantom for EM Evaluation of Handheld Terminals in a Talk Situation," *Proc. Int. IEEE Antennas and Propagation Symp.*, Vol. 2, July 2001, pp. 68–71.

[20] Ogawa, K., H. Iwai, and J. Hatakenaka, "A High-Precision Real Human Phantom for EM Evaluation of Handheld Terminal Antennas in a Talk Situation," *Trans. on IEICE*, Vol. J85-B, No. 5, 2002, pp. 676–686 (in Japanese).

[21] Hall, H., and P. Chadwick, "Assessment of SAR from In-Vehicle Radio Devices Using a Novel Measurement System," *Proc. Joint Meeting of Bioelectromagnetics Society and the European Bioelectromagnetics Association*, (CD-ROM), Dublin, Ireland, June 19–24, 2005, 10-1, pp. 89–90.

[22] Okano, Y., et al., "The SAR Evaluation Method by a Combination of Thermographic Experiments and Biological Tissue-Equivalent Phantoms," *IEEE Trans. on Microwave Theory and Techniques*, Vol. 48, No. 11, 2000, pp. 2094–2103.

[23] Ishido, R., et al., "A Study on the Solid Phantoms for 3–6 GHz and Evaluation of SAR Distributions Based on the Thermographic Method," *Proc. 2004 Int. Symp. Electromagnetic Compatibility*, EMC'04, Sendai, Japan, Vol. 3, B3-2, June 2004, pp. 577–580.

[24] Okano, Y., K. Ito, and H. Kawai, "Solid Phantom Composed of Glycerin and Its Application to SAR Estimation," *Trans. on IEICE*, Vol. J83-B, No. 4, 2000, pp. 534–543 (in Japanese).

[25] Tell, R. A., *Microwave Energy Absorption in Tissue*, EPA, Twinbrook Research Laboratory Technical Report, 1972.

[26] Kritikos, H. N., and H. P. Schwan, "Hot Spot Generated in Condition Spheres by Electromagnetic Waves and Biological Implications," *IEEE Trans. on Biomed. Eng.*, Vol. 19, No. 1, 1972, pp. 53–58.

[27] Shapiro, A. R., R. F. Lutomirski, and H. T. Yura, "Induced Fields and Heating Within a Cranial Structure Irradiated by an Electromagnetic Plane Wave," *IEEE Trans. on Microwave Theory and Techniques*, Vol. 19, No. 2, 1971, pp. 187–196.

[28] Joines, W. T., and R. J. Spiegel, "Resonance Absorption of Microwaves by the Human Skull," *IEEE Trans. on Biomed. Eng.*, Vol. 21, No. 1, 1974, pp. 46–48.

[29] Weil, C. M., "Absorption Characteristics of Multilayered Sphere Models Exposed to UHF/Microwave Radiation," *IEEE Trans. on Biomed. Eng.*, Vol. 22, No. 6, 1975, pp. 468–476.

[30] Massoudi, H., et al., "Electromagnetic Absorption in Multilayered Cylindrical Models of Man," *IEEE Trans. on Microwave Theory and Techniques*, Vol. 27, No. 10, 1979, pp. 825–830.

[31] Nishizawa, S., and O. Hashimoto, "Effective Shielding Analysis for Three Layered Human Model," *IEEE Trans. on Microwave Theory and Techniques*, Vol. 47, No. 3, 1999, pp. 277–283.

[32] Yee, K. S., "Numerical Solution of Initial Boundary Value Problems Involving Maxwell's Equation in Isotropic Media," *IEEE Trans. on Antennas and Propagation*, Vol. 14, No. 3, pp. 1996, 302–307.

[33] Harrington, R. F., *Field Computation by Moment Methods*, New York: Macmillan, 1968.

[34] COST244 Working Group 3, "Proposal Numerical Canonical Models in Mobile Communications," *Proc. COST 244*, Rome, Italy, November 17–19, 1994, pp. 1–7.

[35] Okoniewski, M., and M. A. Stuchly, "A Study of the Handset Antenna and Human Body Interaction," *IEEE Trans. on Microwave Theory and Techniques*, Vol. 44, No. 10, 1996, pp. 1865–1873.

[36] Hombach, V., et al., "The Dependence of EM Energy Absorption Upon Human Head Modeling at 900 MHz," *IEEE Trans. on Microwave Theory and Techniques*, Vol. 44, No. 10, 1996, pp. 1865–1873.

[37] Gandhi, O. P., G. Lazzi, and C. M. Furse, "Electromagnetic Absorption in the Human Head and Neck for Mobile Telephones at 835 MHz and 1900 MHz," *IEEE Trans. on Microwave Theory and Techniques*, Vol. 44, No. 10, 1996, pp. 1884–1897.

[38] Schoenborn, F., M. Burkhardt, and N. Kuster, "Difference in Energy Absorption Between Heads of Adults and Children in the Near Field of Sources," *Health Phys.*, Vol. 74, No. 2, 1998, pp. 160–168.

[39] Wang, J., and O. Fujiwara, "Comparison and Evaluation of Electromagnetic Absorption Characteristics in Realistic Human Head Models of Adult and Children for 900-MHz Mobile Telephones," *IEEE Trans. on Microwave Theory and Techniques*, Vol. 51, No. 3, 2003, pp. 966–971.

[40] Wiart, J., et al., "Children Head RF Exposure Analysis," *Proc. Joint Meeting of Bioelectromagnetics Society and the European Bioelectromagnetics Association*, (CD-ROM), Dublin, Ireland, June 19–24, 2005, 12–8, pp. 146–147.

[41] Dimbylow, P. J., "The Development of Realistic Voxel Phantoms for Electromagnetic Field Dosimetry," *Proc. Int. Workshop on Voxel Phantom Development*, Chilton, United Kingdom, July 6–7, 1995, pp. 1–7.

[42] ICRP, "Report of the Task Group on Reference Man," *ICRP Publication 23*, 1975.

[43] Mason, P. A., et al., "Effects of Frequency, Permittivity, and Voxel Size on Predicted Specific Absorption Rate Values in Biological Tissue During Electromagnetic-Field Exposure," *IEEE Trans. on Microwave Theory and Techniques*, Vol. 48, No. 11, 2000, pp. 2050–2057.

[44] U.S. National Library of Medicine, National Institute of Health, "The Visible Human Project," http://www.nlm.nih.gov/research/visible/visible_human.html, July 2005.

[45] Xu, X. G., T. C. Chao, and A. Bozkurt, "VIP-Man: An Image-Based Whole-Body Adult Male Model Constructed from Color Photographs of the Visible Human Project for Multiparticle Monte Carlo Calculations," *Health Phys.*, Vol. 78, 2000, pp. 476–486.

[46] Nagaoka, T., et al., "Development of Realistic High-Resolution Whole-Body Voxel Models of Japanese Adult Males and Females of Average Height and Weight, and Application of Models to Radio-Frequency Electromagnetic-Field Dosimetry," *Physics in Medicine and Biology*, Vol. 49, 2004, pp. 1–15.

[47] Kainz, W., et al., "Development of Pregnant Woman Models for Nine Gestational Ages and Calculation of Fetus Heating During Magnetic Resonance Imaging (MRI)," *Proc. Joint Meeting of Bioelectromagnetics Society and the European Bioelectromagnetics Association*, (CD-ROM), Dublin, Ireland, June 19–24, 2005, pp. 137–139.

[48] Keller, J. B., "Geometrical Theory of Diffraction," *J. Opt. Soc. Amer.*, Vol. 52, 1962, pp. 116–130.

[49] Bernardi, P., et al., "A UTD/FDTD Investigation on Procedures to Assess Compliance of Cellular Base-Station Antennas with Human-Exposure Limits in a Realistic Urban Environment," *IEEE Trans. on Microwave Theory and Techniques*, Vol. 51, No. 12, December 2003, pp. 2409–2417.

[50] Ghaddar, M., L. Talbi, and T. A. Denidni, "Human Body Modelling for Prediction of Effect of People on Indoor Propagation Channel," *Electronics Letters*, Vol. 40, No. 25, December 9, 2004, pp. 1592–1594.

[51] Ji, Z., et al., "Efficient Ray-Tracing Methods for Propagation Prediction for Indoor Wireless Communication," *IEEE Antennas and Propagation Magazine*, Vol. 43, No. 2, April 2001.

[52] Villanese, F., et al., "Hybrid Image/Ray-Shooting UHF Radio Propagation Predictor for Populated Indoor Environments," *Electronics Letters*, Vol. 35, No. 21, October 14, 1999, pp. 1804–1805.

[53] Wang, J. Q., M. Komatsu, and O. Fujiwara, "Human Exposure Assessment Using a Hybrid Technique Based on Ray-Tracing and FDTD Methods for a Cellular Base-Station Antenna," *Proc. Radio Science Conference, 2004*, Asia-Pacific, August 24–27, 2004, p. 538.

[54] Chen W. T., and H. R. Chuang, "Numerical Computation of Human Interaction with Arbitrarily Oriented Superquadric Loop Antennas in Personal Communications," *IEEE Trans. on Antennas and Propagation*, Vol. 46, No. 6, June 1998, pp. 821–828.

[55] Chen, W.-T., and H.-R. Chuang, "Numerical Computation of the EM Coupling Between a Circular Loop Antenna and a Full-Scale Human-Body Model," *IEEE Trans. on Microwave Theory and Techniques*, Vol. 46, No. 10, Part 1, October 1998, pp. 1516–1520.

[56] Chen, W. -T., and H.-R. Chuang, "Numerical Computation of the EM Coupling Between a Circular Loop Antenna and a Full-Scale Human-Body Model," *IEEE Trans. on Microwave Theory and Techniques*, Vol. 46, No. 10, Part 1, October 1998, pp. 1516–1520.

[57] Meyer, F. J. C., et al., "Human Exposure Assessment in the Near Field of GSM Base-Station Antennas Using a Hybrid Finite Element/Method of Moments Technique," *IEEE Trans. on Biomedical Engineering*, Vol. 50, No. 2, February 2003, pp. 224–233.

[58] Ross, C. T. F., *Advanced Applied Finite Element Method*, Chichester, U.K.: Horwood Publishers, 1998.

[59] Renhart W., et al., "Modelling and Calculation of Influences of RF-Fields on the Human Body Using the Finite Element Method," *IEEE Trans. on Magnetics*, Vol. 30, No. 5, September 1994, pp. 3092–3095.

[60] Schneider, J. B., and K. L. Shlager, "FDTD Simulations of TEM Horns and the Implications for Staircased Representations," *IEEE Trans. on Antennas and Propagation*, Vol. 45, No. 12, December 1997, pp. 1830–1838.

[61] Chen, J.-Y., and O. P. Gandhi, "Currents Induced in an Anatomically Based Model of a Human for Exposure to Vertically Polarized Electromagnetic Pulses," *IEEE Trans. on Microwave Theory and Techniques*, Vol. 39, No. 1, January 1991, pp. 31–39.

[62] Lazzi, G., and O. P. Gandhi, "A Mixed FDTD-Integral Equation Approach for On-Site Safety Assessment in Complex Electromagnetic Environments," *IEEE Trans. on Antennas and Propagation*, Vol. 48, No. 12, December 2000, pp. 1830–1836.

[63] Toftgard, J., S. N. Homsleth, and J. B. Andersen, "Effects on Portable Antennas of the Presence of a Person," *IEEE Trans. on Antennas and Propagation Society*, Vol. 41, No. 6, June 1993, pp. 739–746.

[64] Jensen, M. A., and Y. Rahmat-Samii, "EM Interaction of Handset Antennas and a Human in Personal Communications," *IEEE Proc.*, Vol., 83, No. 1, January 1995, pp. 7–17.

[65] Furse, C. M., and O. P. Gandhi, "The Use of the Frequency-Dependent Finite-Difference Time-Domain Method for Induced Current and SAR Calculations for a Heterogeneous Model of the Human Body," *IEEE Trans. on Electromagnetic Compatibility*, Vol. 36, No. 2, May 1994, pp. 128–133.

[66] Dimbylow, P. J., and S. M. Mann, "SAR Calculations in an Anatomically Realistic Model of the Head for Mobile Communications Transceivers at 900 MHz and 1.8 GHz," *Physics in Medicine and Biology*, Vol. 39, No. 10, 1994, pp. 1537–1553.

[67] Luebbers, R., and R. Baurle, "FDTD Predictions of Electromagnetic Fields In and Near Human Bodies Using Visible Human Project Anatomical Scans," *Antennas and Propagation Society Int. Symp.*, 1996 AP-S Digest, Vol. 3, July 21–26, 1996, pp. 1806–1809.

[68] Sarkar T. K., et al., "A Survey of Various Propagation Models for Mobile Communication," *Antennas and Propagation Magazine*, Vol. 45, No. 3, June 2003.

[69] Wang, Y., S. Safavi-Naeini, and S. K. Chaudhuri, "A Hybrid Technique Based on Combining Ray Tracing and FDTD Methods for Site-Specific Modelling of Indoor Radio Wave Propagation," *IEEE Trans. on Antennas and Propagation*, Vol. AP-48, No. 5, May 2000, pp. 743–754.

[70] Attiya, M., and A. Safaai-Jazi, "Simulation of Ultra-Wideband Indoor Propagation," *Microwave and Optical Technology Letters*, Vol. 42, No. 2, July 20, 2004.

[71] Zhao, Y., Y. Hao, and C. G. Parini, "Two Novel FDTD Based UWB Indoor Propagation Models," *2005 IEEE Int. Conference on Ultra-Wideband (ICU 2005)*, Zurich, Switzerland, September 5–8, 2005.

[72] Hao, Y., and C. J. Railton, "Analyzing Electromagnetic Structures with Curved Boundaries on Cartesian FDTD Meshes," *IEEE Trans. on Microwave Theory and Techniques*, Vol. 46, January 1998, pp. 82–88.

[73] Holland, R., "Finite Difference Solutions of Maxwell's Equations in Generalized Nonorthogonal Coordinates," *IEEE Trans. on Nuc. Sci.*, Vol. NS-30, No. 6, December 1983, pp. 4689–4591.

[74] Hao, Y., "The Development and Characterization of a Conformal FDTD Method for Oblique Electromagnetic Structures," Ph.D. thesis, University of Bristol, United Kingdom, November 1998.

[75] Hao, Y., et al., "Numerical Modeling of On-Body Radio Propagation Channel," *Proc. of IEEE AP-S*, Washington, D.C., July 3–8, 2005.

[76] Weedon, W. H., and C. M. Rappaport, "A General Method for FDTD Modeling of Wave Propagation in Arbitrary Frequency Dispersive Media," *IEEE Trans. on Antennas and Propagation*, Vol. 45, No. 3, March 1997.

CHAPTER 3

Antennas and Propagation for On-Body Communication at Microwave Frequencies

Peter S. Hall, Yang Hao, Yuriy Nechayev, Akram Alomainy, Costas Constantinou, Clive Parini, Muhammad Kamarudin, and Tareq Salim

3.1 Introduction

The preface to this book and Chapter 1 have emphasized that on-body communications systems will take their place as one of the dimensions of personal and body area networks, and that this is an area of increasing interest. Body area networks can be used to interconnect the various components in a wearable computer system. This system may be supporting a body sensor network for health monitoring or drug release, for personal entertainment or information delivery, or for occupational support. While these systems may use wires for their interconnection, whether untethered or built into a smart fabric, there is now a trend towards wireless interconnection. An example of this is the cellphone Bluetooth headsets now widely available.

However, there appears to be little published on the topic in the open literature that specifically addresses the electromagnetic components of on-body communication systems. There are many publications on the effect of the human body on the operation of antennas located in close proximity [1–4]; on the absorption of energy within the body, and in particular, SAR for proximate antennas [5–7]; and of the propagation on and off the body for use in cellphone and pager systems [8, 9]. Therefore, many of the tools needed for the study of on-body channels already exist. However, there are some significant differences in approach.

In conventional mobile communications systems, between a mobile base station and a mobile terminal, variation in the channel is due to the interference between multiple rays scattered from the local environment, such as buildings in the outdoor case, and walls or furniture in the indoor case. While local scattering will play a part in communications between two terminals mounted on the body, significant variations in the channel are also due to changes in the geometry of the body. Even when standing or sitting, the body is subject to many small movements. During normal activities, movement becomes significant, and during the playing of sport, movement may be extreme. Thus, characterization of radiowave propagation needs to account both for the variable positioning of the terminals on the body, and for dramatic changes in the geometry of the local environment. Second, the local geometry changes may also affect the operation of the antenna, in particular the input match and radiation pattern. Changes in the antenna and the propagation loss are operational considerations that must be understood during the design of the transceivers, to enable maximum channel capacity and minimum power consumption.

Figure 3.1 shows possible transmitter or receiver locations on the body, which thus indicate the channels, or paths, whose characteristics are discussed below. It is clear that there are potentially many paths in an on-body network. Communications between on-body modules will involve paths on the body, as well as those scattering off the local environment of the body, whether they are indoors or outdoors. Link loss has been measured both in an anechoic chamber and in other surroundings, such as a laboratory, and significant differences have been noticed. There are now models available for both indoor and outdoor propagation. The challenge is to combine the models developed for the on-body propagation with existing local environmental models or statistics.

In addition, during the channel characterization phase, there are difficulties in de-embedding the antenna characteristics from those of the propagation path. Figure 3.2 shows a body with an antenna mounted on the shoulder and on the belt. The ground is ignored for simplicity. In the case of communications on and off the body, to both local and distant base stations, it is appropriate to characterize the antenna radiation pattern at some distance from the body. Indeed, some handset antennas are measured in this way by placing both the body and antenna in the anechoic chamber. However, for the shoulder-to-belt link indicated, this type of characterization is certainly not appropriate, and a characterization of the radiation fields local to the antenna, including those in the body tissue close to the surface, need to be determined to find the coupling of the antenna to the propagation path. In the following sections, this problem is not fully discussed. Rather, the pragmatic approach

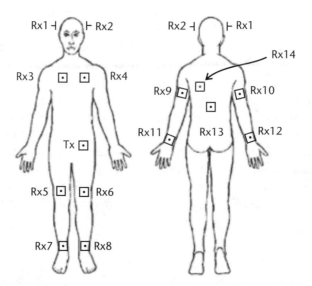

Figure 3.1 Possible transceiver locations on the body. (*From:* [10]. © 2005 IEEE. Reprinted with permission.)

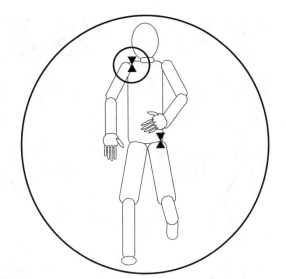

Figure 3.2 Range at which to specify antenna radiation pattern. For on-body link, specify close to antenna, shown by small circle; for off-body link, specify outside region of body, shown by large circle. (*From:* [10]. © 2005 IEEE. Reprinted with permission.)

of characterizing the channel, using what could be called a standard antenna, is used. However, a first approach to the de-embedding problem is to simulate the far-field radiation patterns of the antenna on a suitable numerical phantom. For example, such a simulation of a monopole and planar inverted-F antenna in free space and on a cylindrical shaped body phantom at 2.45 GHz shows differences of around 5 dB. Differences of more than 5 dB are noticed for both antennas. Full de-embedding would require computation of the coupling factor between the body mounted antenna and the local suface or creeping ray.

The choice of frequency is not straightforward. There are examples of on-body communications at frequencies below the microwave bands, and these are discussed in detail in Chapter 4. However, the significant increase in the use of Bluetooth or WLAN modules for wearable computers has led us to focus initial efforts on an unlicensed industrial, scientific, and medical (ISM) band. Thus, all the measurements described in this chapter have been made at 2.45 GHz.

The University of Birmingham and Queen Mary and the University of London have been characterizing on-body radiowave propagation and antennas for some time [10–18]. The following sections describe the results.

3.2 On-Body Channel Measurement and Modeling

The propagation path loss of an on-body channel was measured using a vector network analyzer (VNA) inside an anechoic chamber. The antennas used were quarter-wavelength monopoles located over small ground planes. They were attached to the body at a number of positions and orientations, and connected to the calibrated VNA by two 5-m-long flexible coaxial cables. Figure 3.3 shows an example of the attachment arrangements. For each antenna placement setup, the S_{21} response was

Figure 3.3   Placement of test antennas on the body (chest-to-belt link).

then measured every 0.225 second while the person wearing the antennas changed body positions every 20 seconds. The positions of the antennas on the body are shown in Figure 3.1, and the body positions, or postures, used in all the measurements are listed in the caption to Figure 3.4. Antenna and cable mountings were chosen so as not to impede body movement.

On-body propagation links can be roughly categorized according to the parts of the body at which the transmit and receive antennas are attached (e.g., trunk-to-trunk, trunk-to-head, and trunk-to-limb). Characteristics of the propagation channels corresponding to different types of such links can be expected to differ considerably, due to link geometry variability. For example, a trunk-to-limb link is expected to be subject to significant variation due to the movement of the limb, while a trunk-to-trunk link will be more stable. The receiver positions in Figure 3.4 were chosen to illustrate this, with a belt-to-chest link likely to experience the least perturbation in the geometry of the link, while the belt-to-head will experience more, and the belt-to-wrist will experience the most.

The measurements confirm the simple hypothesis. The belt-to-chest link, being the shortest, has the lowest loss, with an average loss of 41 dB, and has a

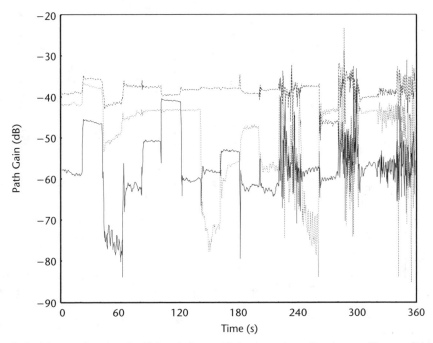

Figure 3.4   Measured path gain (S_{21}) variations with body posture. Receiver positions: solid = right side head (Rx1, Figure 3.1); chain = right chest (Rx3); dotted = right wrist (Rx12). Transmitter on belt, left side, frequency = 2.45 GHz, measured with quarter-wavelength monopoles; size = 25 × 2 mm copper wire; ground plane = 80 mm diameter, 0.4-mm-thick copper sheet, mounted 1 cm away from body. Postures: 0–20 standing upright; 20–40 standing, body turned left; 40–60 standing, body turned right; 60–80 standing, body leaned forward; 80–100 standing, head leaned forward; 100–120 standing, head turned left; 120–140 standing, head turned right; 140–160 standing, arms stretched out to sides; 160–180 standing, arms above head; 180–200 standing, arms reaching forward; 200–220 standing, forearms forward; 220–240 standing, moving arms, head, and body freely; 240–260 sitting, arms hanging body; 260–280 sitting, hands in lap; 280–300 sitting, moving arms, head, and body freely; 300–320 standing; 320–340 walking back and forth, arms to body; 340–360 walking, moving arms, head, and body freely.

peak-to-peak variation of approximately 8 dB in the stationary postures, which are those up to 220 seconds, and from 240 to 280 seconds. During the moving postures (from 220 to 240 seconds, from 280 to 300 seconds, and from 320 to 360 seconds), the variation is greater, at 21 dB. The belt-to-head link has a higher average loss, primarily due to the longer link length. However, the variations in the stationary postures are now much greater, at 44 dB, during the stationary postures, and 56 dB in the movements. The belt-to-wrist link has losses of 41 and 59 dB during stationary postures and movement, respectively. It can also be seen from Figure 3.4 that variations within the stationary posture periods, due to breathing and/or other small involuntary body movements, are no more than approximately 4 dB, which agrees with the results reported in [11]. The average path gains for each body posture and antenna position are given in Table 3.1.

The results of Figure 3.4 and Table 3.1 are taken with the human body inside the anechoic chamber, standing on a reinforced concrete floor. Significant differences are seen when the body is located in a conventional environment. Table 3.2

Table 3.1 Mean Path Gain, in Decibels

Posture	Chest Right	Chest Right (rpt)	Head Right	Wrist Right	Back Center	Chest Left	Head Left	Head Left (rpt)	Left Wrist	Right Ankle	Left Ankle	Left Thigh	Left Thigh (rpt)	Right Thigh	Left Upper Arm	Right Upper Arm
Stand	−40.4	−39.1	−58.7	−41.9	−49.4	−36.5	−44.9	−44.6	−29.2	−46.1	−44.3	−36.9	−36.1	−40.4	−41.3	−50.8
Body Left	−36.7	−35.7	−46.2	−37.5	−51.5	−34.9	−50.3	−44.5	−31.3	−50.4	−46.4	−37.0	−35.7	−41.1	−45.8	−46.0
Body Right	−41.8	−41.8	−75.3	−50.5	−46.3	−35.8	−44.4	−42.6	−26.9	−42.8	−42.9	−34.2	−33.3	−32.9	−37.4	−60.2
Body Forward	−37.6	−37.4	−62.4	−44.8	−47.8	−31.6	−41.4	−45.5	−30.9	−49.4	−45.3	−34.7	−33.7	−35.5	−39.7	−53.1
Head Forward	−38.7	−37.7	−50.9	−43.5	−50.0	−37.6	−38.6	−39.6	−28.8	−47.6	−43.5	−37.7	−36.3	−40.4	−41.3	−52.1
Head Left	−40.1	−39.6	−41.0	−43.3	−49.7	−37.3	−63.0	−64.1	−28.6	−48.1	−43.7	−37.9	−36.7	−40.3	−41.2	−51.5
Head Right	−39.3	−38.3	−60.5	−43.3	−49.8	−37.1	−41.6	−42.3	−28.6	−47.7	−43.5	−38.0	−36.8	−39.9	−40.4	−51.3
Arms Sideways	−38.2	−37.9	−58.3	−72.7	−52.0	−35.2	−41.6	−44.1	−64.7	−47.8	−43.1		−36.6	−40.0	−39.8	−58.5
Arms Upwards	−36.7	−38.0	−53.3	−57.5	−51.6	−36.4	−43.4	−43.3	−50.5	−48.2	−43.7		−36.4	−39.7	−46.2	−59.6
Arms Forward	−40.6	−39.2	−62.1	−47.7	−50.9	−32.3	−52.1	−53.7	−37.8	−50.6	−43.4		−36.9	−37.4	−35.5	−53.1
Forearms Forward	−38.6	−38.4	−62.9	−57.4	−48.7	−42.9	−57.7	−60.8	−31.7	−46.6	−42.8		−36.5	−38.3	−47.7	−72.6
Stand Free	−39.2	−39.2	−55.4	−55.9	−52.7	−35.5	−48.3	−47.4	−42.8	−47.0	−43.3	−38.0	−36.2	−38.9	−45.9	−60.6
Sit	−39.8	−37.5	−58.2	−70.7	−50.1	−29.2	−46.2	−43.3	−33.6	−61.1	−59.1	−34.4	−33.7	−36.8	−43.1	−48.7
Sit Hands in Lap	−38.6	−46.3	−60.0	−43.9	−49.8	−29.5	−41.7	−42.2	−37.3	−66.4	−63.7	−38.8	−37.9	−40.8	−42.6	−49.7
Sit Free	−37.7	−36.8	−52.4	−51.3	−50.4	−32.7	−42.3	−42.8	−44.3	−64.1	−58.0	−35.1	−37.0	−38.8	−43.4	−58.7
Stand	−37.6	−40.2	−57.4	−43.8	−48.6	−37.1	−44.4	−44.8	−30.2	−48.3	−41.2	−36.6	−35.5	−39.4	−44.3	−50.2
Walk	−39.9	−38.9	−57.5	−44.4	−49.2	−36.2	−45.3	−43.5	−30.6	−51.7	−41.0	−34.5	−34.1	−40.0	−43.9	−53.5
Walk Free		−37.7	−56.2	−51.1	−51.3	−36.5	−44.3	−44.3	−44.1	−51.9	−40.8	−34.7		−39.9	−46.2	

For paths shown in Figure 3.1, Tx on belt, left side, measured with quarter-wave monopoles; frequency = 2.45 GHz.
Source: [10].

Table 3.2 Mean Path Gain and Path Gain Variability (Range)

Rx Position		Path Gain (dB)			
		Anechoic Chamber		Laboratory	
		Mean	Range	Mean	Range
Trunk	Rx3(chest)	−41	14	−44	24
	Rx14(back)	−60	20	−57.5	25
Head	Rx2(left side)	−52	36	−40.5	29
	Rx1(right side)	−53.5	33	−41	38
Wrist	Rx11(left)	−46.5	45	−51.5	37
	Rx12(right)	−55.5	29	−57	26

For posture shown in Figure 3.4 and paths shown in Figure 3.1, Tx on belt, left side; measured with quarter-wave monopoles; frequency = 2.45 GHz.

shows mean path gain and variability, both in the chamber and in the laboratory. In the trunk-to-trunk links, the mean losses are similar, but the variability is increased by up to 10 dB in the laboratory. For trunk-to-head links, the mean path gain reduces by more than 10 dB, while the variability is either increased or decreased. For trunk-to-wrist links, the loss increases while the variability decreases. It is concluded the local environment does have an effect on the path loss, and combined modeling of both mechanisms is required.

The results show that the major source of variability in the channel is movement of the parts of the body to which the antennas are attached. These movements cause variations in the distance between the antennas. The channel variability is especially severe when one of the antennas is positioned on a hand, which is highly mobile. The highly variable dependence of the mean path gain with antenna separation distance is demonstrated in Figure 3.5. The distance was measured along the shortest path between the antennas, which does not penetrate the body. The straight line corresponding to a simple power law fit for the magnitude of the path gain is also shown. It gives an estimate of the signal attenuation with the distance, even though shadowing and polarization misalignment also contributes to the variations. It can be seen that the path loss exponent equals 3.2, which is in good agreement with the result reported in [19]. However, it is clear that the variability of the path gain about such a simple power law fit is great enough to prevent predictions irrespective of additional antenna position information.

The probability distribution of the signal strength, as shown in Figure 3.6, was obtained from the data from the moving postures in Figure 3.4, which are from 220 to 240 seconds, from 280 to 300 seconds, and from 320 to 360 seconds. The data from these postures was combined for each receiver position, which are right chest (Rx3), right ankle (Rx7), and right wrist (Rx12), in order to obtain probability distributions of the path gain. The distributions for the other receiver positions are similar. With the exceptions of ankle-mounted antenna positions, the probability distributions are approximately log-normal for positions Rx3 and Rx12. The spread of the distribution strongly depends on the degree of freedom of movement that exists for the transmitter and the receiver; therefore, when the antenna is mounted on the chest (position Rx3), the distribution is noticeably narrower than when the antenna is on the wrist (position Rx12). On the other hand, for the

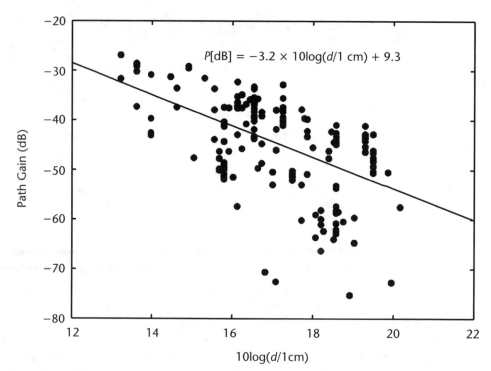

$$P[dB] = -3.2 \times 10\log(d/1\ cm) + 9.3$$

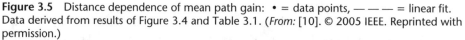

Figure 3.5   Distance dependence of mean path gain: • = data points, — — — — = linear fit. Data derived from results of Figure 3.4 and Table 3.1. (*From:* [10]. © 2005 IEEE. Reprinted with permission.)

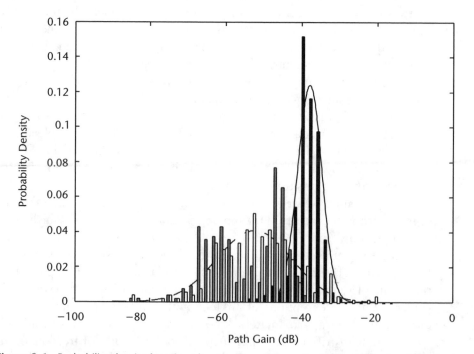

Figure 3.6   Probability density functions for moving postures. Rx positions: black = right chest (Rx3); gray = right ankle (Rx7); white = right wrist (Rx12). Tx on belt, left side. Data in histogram bars, lines are Gaussian fits, for Rx3 and Rx12. Data derived from results of Figure 3.4 and Table 3.1. (*From:* [10]. © 2005 IEEE. Reprinted with permission.)

ankle-mounted antenna positions (such as position Rx7 in Figure 3.6), two peaks are observed in the probability distribution. They correspond to movements in standing and sitting positions having two different average distances between the antennas.

Variation of the signal strength around a human torso has also been investigated. Two quarter-wavelength monopoles were mounted on the waist, as shown in Figure 3.7(a). The receiving antenna was moved away from the transmitter, and the path gain was measured every 2 cm. The path gain expressed in decibels, as shown in Figure 3.7(b), can be seen to fall off linearly with the distance, corresponding to an exponential attenuation of the signal, which is characteristic of a creeping wave. Straight lines were fitted to the data as shown, with the crossover point between the two curves being approximately diagonally opposite the excitation point. The right-hand ordinate is close to the excitation antenna, and approximately corre-

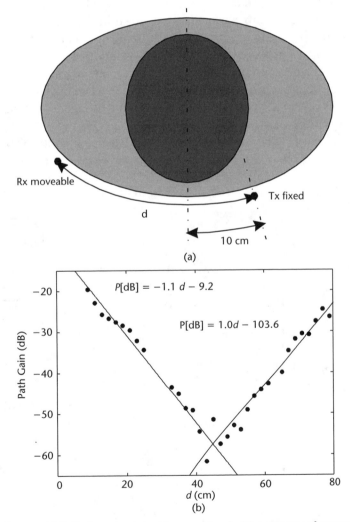

Figure 3.7 Measured falloff of received signal around trunk: (a) geometry of measurement, and (b) results. • = data, lines are linear fits. Measured with quarter-wavelength monopole, as in Figure 3.4, trunk circumference = 81 cm. (*From:* [10]. © 2005 IEEE. Reprinted with permission.)

sponds to 360° around the body. The rate of attenuation, found by fitting straight lines to the curve, is approximately 1 dB/cm.

Figures 3.8 to 3.10 show similar results, at 400 MHz, 900 MHz, and 2.4 GHz, respectively, from Ryckaert et al. [20]. The measurements were taken with dipoles oriented vertically to the body surface, and were taken over an angle of 180° around the body. The propagation mode is assumed to be a creeping wave, which exhibits an exponential decay of power [21]. Beyond the breakpoint angle, the two rays are assumed to interfere, resulting in smaller decay but increased variability. The proposed path loss model is

$$P_{dB}(\theta) = P_{dB}(\theta_0) - \gamma_1(\theta - \theta_0) \quad \text{for} \quad \theta_0 \langle \theta \langle \theta_{bp}$$
$$P_{dB}(\theta) = P_{dB}(\theta_{bp}) - \gamma_2(\theta - \theta_{bp}) \quad \text{for} \quad \theta_{bp} \langle \theta \langle \pi$$

$$(3.1)$$

where γ_1 and γ_2 are the decay coefficients before and after the breakpoint, θ_{bp} is the breakpoint angle, and σ_1 and σ_2 are the standard deviations of the measured data. Values for the various parameters are given in the figures. It is clear that the path losses and the variability increase with frequency. The decay coefficient at 2.4 GHz corresponds approximately to 2 dB/cm, assuming that the body radius is 20 cm. It is clear that significant interference takes place beyond the breakpoint at 400 and 900 MHz, although it is difficult to see at 2.4 GHz. The high level of variability in these figures is due to the combining of data from a number of measurements at different heights on the chest.

Hall et al. [11] have also investigated the variability of path loss around the chest due to breathing. Breathing causes the chest cavity to expand and contract,

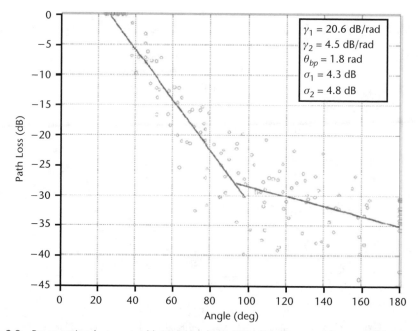

Figure 3.8 Propagation loss around human body at 400 MHz. γ_1, γ_2 = decay coefficients; θ_{bp} = breakpoint angle; σ_1, σ_2 = standard deviations; subscripts refer to before and after breakpoints. (*From:* [20]. © 2004 IEE. Reprinted with permission.)

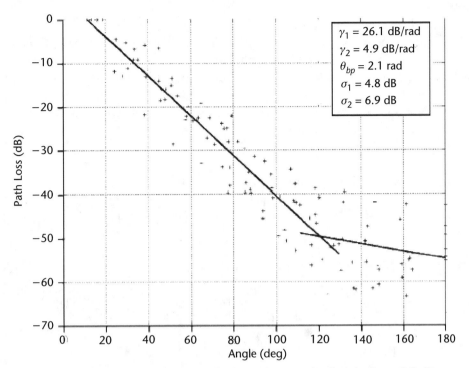

Figure 3.9   Propagation loss around human body at 900 MHz, details as in Figure 3.8. (*From:* [20]. © 2004 IEE. Reprinted with permission.)

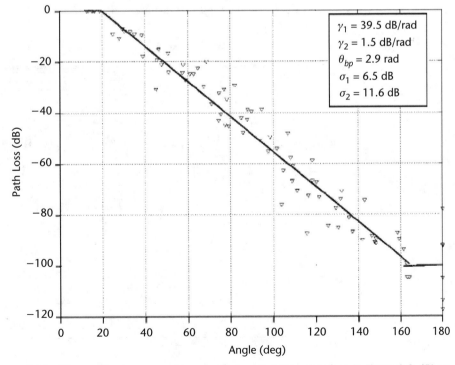

Figure 3.10   Propagation loss around human body at 2.4 GHz, details as in Figure 3.8. (*From:* [20]. © 2004 IEE. Reprinted with permission.)

and this will change the interference between the two creeping rays. Figure 3.11 shows the measured extension of the chest circumference over a period of approximately 240 seconds, with the body stationary. It can be seen that variations of nearly 20 cm occur. This data is used in a ray model to compute the path loss due to two interfering rays, propagating around an elliptical cylinder of size 300 × 200 mm. In the result shown in Figure 3.12, the transmitter and receiver are located 159° apart

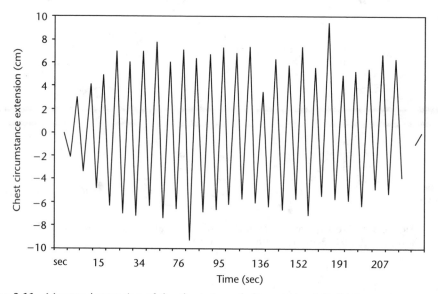

Figure 3.11   Measured extension of the chest circumference during breathing.

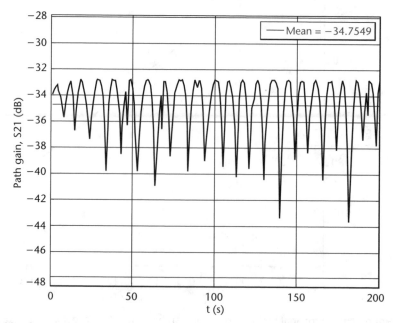

Figure 3.12   Computed path loss between antennas on chest using two-ray model. Tx and Rx 159° apart horizontally around chest, chest extension data as in Figure 3.11.

around the body. Signal variations of approximately 14 dB are observed, around a mean path loss of −60 dB. These are comparable to the variations noted in Figure 3.10, but larger than those noted in Figure 3.4.

The time delays of on-body propagation paths are not easy to measure for narrowband channels. However, they can be estimated from the dimensions of the body. Path lengths on the body do not usually exceed approximately 2m, depending on the antenna positions, which leads to a maximum time delay of 7 ns. However, it is evident from the measured path gains that the environment around the body can significantly contribute to the received signal. Therefore, the paths scattered from the surrounding objects will increase the delay spread. Inside a room, the singly-scattered path lengths are on the order of twice the room dimensions. For example, a 5-m-long room gives an estimated time delay $\tau = 33$ ns, corresponding to the coherence bandwidth $B_c \cong 1/\tau = 30$ MHz. For a more accurate prediction, direct measurements of the time delay are required.

3.3 Antenna Design

There are two primary requirements for antennas for on-body links. First, the antenna needs to be insensitive to the proximity to the body; and second, the antenna needs to have a radiation pattern shape that minimizes the link loss. The matching of the antennas is naturally affected by their proximity to the body, due to the influence of the body on the antenna reactive fields. A number of techniques are available [22, 23] for adaptively matching antennas, but a discussion of these is outside the scope of this chapter. Instead, we focus on the antenna radiation patterns and their influence on the observed path gain.

Such measurements show that the path gain is highly dependent both on antenna type and link geometry. The antenna positions also determine the propagation mode. For example, for the belt-to-chest path, it is evident that propagation will be predominantly due to a surface or creeping wave. Similarly, the belt-to-wrist path will be in some cases a free-space path, when the hand is in front of the body; and in others, it will be a shadowed free-space path with diffraction around the body, when the arm is behind the body. In the case of a grazing wave around the surface, it is clear that an antenna with a radiation pattern peak along the body surface and with vertical polarization is needed. For a free-space link, an antenna with maximum radiation away from the body is needed. In all cases, radiation into the body should be minimized, since the relatively small skin depth of a few centimeters at these frequencies suggests that little energy is propagated through the body.

Due to the large number of possible link geometries and body postures, it is rather difficult to specify ideal antenna characteristics. Therefore, in this section, results are shown that enable estimates to be made of the best antenna type, and for the number of body postures that will give optimized results. A full statistical description of link performance for the various antennas would require many more measurements than are present here.

3.3.1 Comparison of Antenna Types

Path gain has been measured by placing a variety of antennas at various points on the body, as shown in Figure 3.1. The antennas used were as follows:

- Monopole: 31-mm-high, 0.5-mm-thick copper wire, 50-mm-diameter ground, 0.4-mm-thick copper;
- Loop: 4 turns, 19-mm diameter, 30-mm length, mounted with its axis normal to the body surface;
- Patch: 38×29 mm^2, on 1.6-mm thick FR4 substrate of $\varepsilon_r = 4.4$, and 48×48 mm^2 ground plane;
- Patch array: 2-element patch arrays with phasing to give a broadside beam and beams inclined by 45° and 90°, respectively.

The rationale behind these choices is as follows. The loop and monopole give omnidirectional patterns with maximum radiation along the surface, but with H and E fields normal to the surface, respectively. Such patterns will be suitable for paths where the surface wave dominates. However, the monopole has a polarization that matches that of the surface wave, where the loop is orthogonal. The patch gives strong radiation away from the surface, suitable for the free-space links. The arrays with various beam tilts were chosen, since it is evident that in some circumstances, the paths in the belt-to-wrist case are not normal to the body surface.

Figure 3.13(a) shows measured link loss for the belt-to-chest path for various antennas. It can be seen that one antenna is always a monopole, while the other is changed. The loss varies by approximately 1 to 2 dB within each 20-second period, but this is averaged in the figure. The posture changes are the same as those given in Figure 3.4. It is clear that for all postures, the monopole-monopole combination gives lowest link loss, as expected, due to the orientation of the field polarization relative to the body. Figure 3.13(b) shows the link loss for the belt-to-wrist case, with the same conditions. It can be seen that the variation of loss is much greater than in the belt-to-chest case, and for several postures, the array with beam tilt gives better results.

3.3.2 Antenna Match and Efficiency

The input impedance and efficiency of the antenna will be affected by the proximity of the body. In the design of the antenna, it is important in many body links to excite a surface wave, and this is particularly true in trunk-to-trunk links. This implies strong fields on the antenna ground plane. If the equipment demands the use of a small ground plane, then the reactive near fields of the antenna may interact with the body and cause losses.

Figure 3.14 shows measurements of the input return loss of a monopole as it is brought close to the human body. It can be seen that the resonant frequency has shifted by 300 MHz, which is greater than its −10-dB bandwidth.

The efficiency of a monopole and planar inverted-F antenna on a small ground plane have been measured both in free space and located close to the human body, and the results are shown in Table 3.3. The body is represented by a phantom, as shown in Figure 3.15(a). The measurements in the chamber consist of first comput-

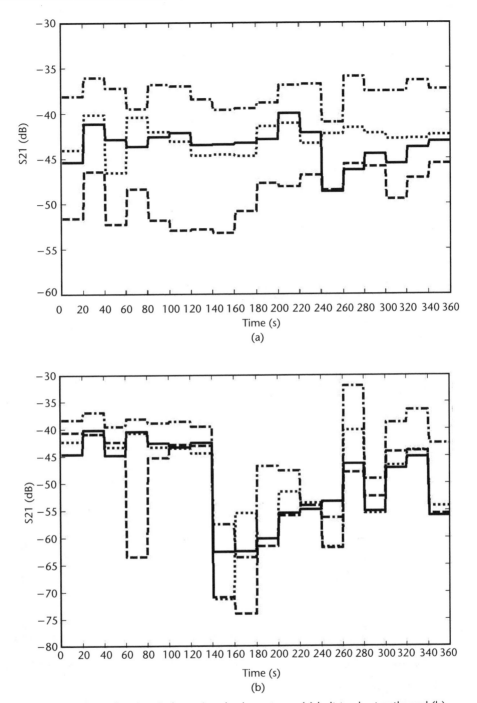

Figure 3.13 Measured path gain for various body postures: (a) belt-to-chest path, and (b) belt-to-wrist path. Frequency = 2.45 GHz; loss averaged over 20-second posture period; monopole-patch = solid; monopole-array = dash; monopole-loop = dot; monopole-monopole = dot-dash. Postures: 0–20 standing; 20–40 standing, body turned left; 40–60 standing, body turned right; 60–80 standing, body bent forward; 80–100 standing, head bent forward; 100–120 standing, head turned left; 120–140 standing, head turned right; 140–160 standing, arms stretched out to side; 160–180 standing, arms above head; 180–200 standing, arms forward; 200–220 standing, forearms forward; 220–240 standing freely*; 240–260 sitting, arms hanging; 260–280 sitting, hands on lap; 280–300 sitting freely*; 300–320 standing; 320–340, walking, arms close to body; 340–360, walking, arms swinging free. *Freely means arbitrary position. (*From:* [18]. © 2005 IEEE. Reprinted with permission.)

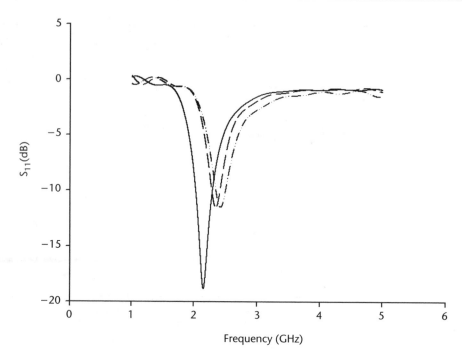

Figure 3.14 Measured input return loss of monopole antenna in proximity to human body phantom. Distance between antenna and phantom, *d*; chain line *d* = 0; dotted chain line *d* = 5 mm; long chain line *d* = 10 mm; dotted line *d* = large, no phantom near antenna; monopole: 31 × 0.5 mm copper wire; ground plane: 50-mm diameter, 0.4-mm-thick copper sheet; phantom: relative permittivity = 40; and conductivity = 2.1 S/m, 20 cm diameter, 5 cm thick.

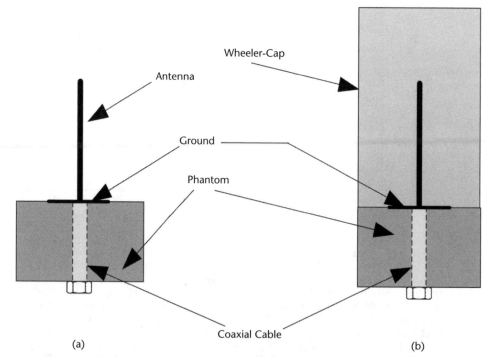

Figure 3.15 Monopole antenna on body phantom: (a) in free space, and (b) in Wheeler cap. Phantom: relative permittivity = 40, and conductivity = 2.1 S/m, 10 cm diameter, 5 cm thick.

ing the directivity from an integration of the measured radiation pattern, and then comparing this to the measured gain. The measurements using the Wheeler cap involve placing the antenna and the phantom in the cap, as shown in Figure 3.15(b). The Wheeler cap was 10 cm in diameter, and the 5-cm-thick phantom completely filled the cap below the antenna. Table 3.3 shows that the two measurements agree well for both antennas. In free space, the efficiency is on the order of 90%. In this case, there are significant currents flowing over the ground plane and onto the coaxial feed cable. These currents are absorbed by small ferrite beads on the cable, giving rise to the small loss. The measured efficiency when mounted on the phantom, where the antenna ground planes are in contact with the phantom material, is between 40% and 50%. There are some detailed differences between the measured results for the monopole and the planar inverted-F antenna (PIFA), but the conclusions are the same. Simulations of the efficiency have also been performed and are shown in Table 3.3. There is agreement within ±5%, but the reduction of the off-resonance efficiency is not predicted well in the case of the monopole. While in practice, the antenna is not likely to be in close contact with the body, and therefore will not experience such losses as shown here, these results serve to show the magnitude of the effect.

Optimized Antenna Design

It is clear that a quarter-wavelength monopole is a suitable type for many on-body links, since it has both an omnidirectional pattern, which obviates the need for correct orientation, and a polarization vertical to the body surface. However, this monopole is probably too big for practical implementation. Figure 3.16 shows a reduced-height monopole formed by top-loading the monopole with a disc. The radiation pattern in the elevation plane of this antenna compared to the full-height quarter-wavelength monopole is shown in Figure 3.17. There are some differences in the pattern when the monopole is mounted on the body. The patterns are each normalized to 0 dB, and thus do not include the effect of reduced efficiency noted above. The performance of the antennas on the body in terms of link loss is shown in Figure 3.18. The postures are those of Figure 3.4, and results are shown for the monopole and the disc-loaded monopole. The results show that the average loss is lower for the disc monopole antenna. However, there is some experimental variation in the repeatability of the mounting of the antennas on the body. Table 3.4 gives the means and standard deviations for several sets of measurements. The dif-

Table 3.3 Efficiency of Antennas on the Body

Efficiency (%)		Monopole		PIFA	
		Free Space	On Phantom	Free Space	On Phantom
Measurement	Pattern integration/ substitution	90	52	90	48
	Wheeler cap	92	44	91	37
Simulation		87	53	92	38

Monopole 31 × 0.5 mm copper wire; ground plane diameter = 50 mm; 0.4-mm thick copper; PIFA, 25.9 × 14.8 mm^2; h = 9 mm; air spaced; ground plane, 50 × 40 mm^2; frequency = 2.45 GHz; phantom as in Figure 3.15.

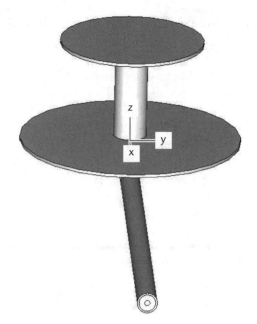

Figure 3.16 Disc-loaded, reduced-height monopole. Disc = 13 × 0.55 mm²; monopole = 17 × 3 mm diameter; ground plane 20 × 0.55 mm copper sheet; and frequency = 2.45 GHz.

ferences between the results for various postures in two datasets taken on different days and averaged to give the mean, while the standard deviation is over the differences for the various postures. It can be deduced from these results that the performance of the disc and monopole antennas is similar within the experimental error.

3.4   Simulation and Modeling

Since radio propagation on and around the human body, even for a single static posture, is a complex electromagnetic problem, numerical simulation tools can provide physical insight into the propagation mechanisms; support the extraction of simple closed form equations for path gain; and enable extension into modeling of the body in the local environment.

A conformal FDTD modeling software, developed at Queen Mary, University of London, was applied to the trunk-to-trunk case, with a monopole used as the transmit antenna and probes used to find the electric field strength near the source and at certain distances from the chest. For dry skin, body parameters used were relative permittivity equal to 38.1 and conductivity equal to 1.441 S/m. For muscle, which is used over the whole body, the parameters were 52.7 and 1.705 S/m, respectively. For lungs, which are a large volume in the upper body, the parameters are 34.5 and 1.219 S/m, respectively. Bones and other organs are not considered because of their relatively small volumes. The frequency was 2.4 GHz. The dimensions of the semimajor and semiminor trunk axes are 150 and 120 mm, respectively. The lungs were approximately 160 × 130 mm, and the thickness of the skin was 2 mm.

To simulate path loss, a monopole antenna is used to excite a wave, and two probes were used to sample the power reduction along the path. One cell outside the

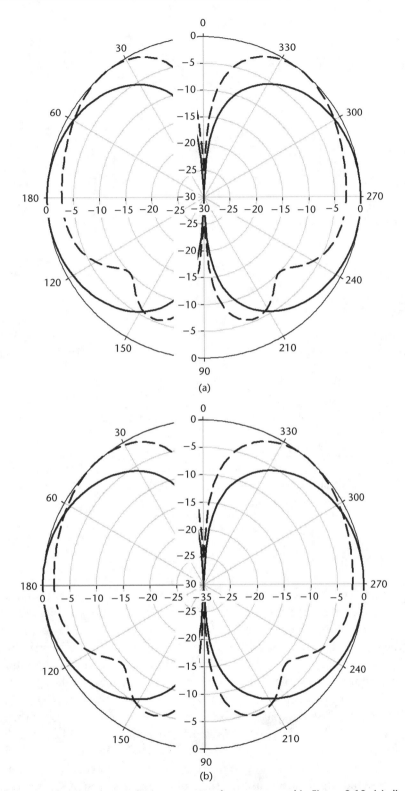

Figure 3.17   Simulated E-plane radiation patterns of antennas used in Figure 3.18: (a) disc-loaded monopole, and (b) quarter-wavelength monopole. Solid = in free space; dotted = in contact with body phantom. Quarter-wavelength monopole details as in Figure 3.4. Disc-loaded monopole details as in Figure 3.16. Phantom as in Figure 3.14. Patterns are normalized to 0 dB.

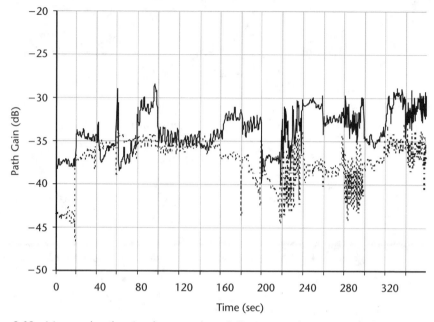

Figure 3.18 Measured path gain of monopole and disc monopole antennas on human body. Solid = mono-disc; dotted = mono-mono. Belt-to-chest path; other details as in Figure 3.13.

Table 3.4 Mean Difference and Standard Deviations of Repeated On-Body Measurements

Receiver Position	Right Chest	Left Head	Left Thigh
	Rx3	Rx2	Rx6
Mean difference (dB)	1.3	1.7	1.1
Standard deviations (dB)	1.77	1.56	0.41

Note: Transmitter is on the belt.

body next to the skin around the waist was chosen as the monopole source. The ground plane of the monopole was 60 × 60 mm, and was located on the surface of the body. Probe 1 was placed 7 cm away from the source; Probe 2 was positioned on the chest, 28.6 cm away from Probe 1. All time-domain electric field components collected at Probes 1 and 2 were converted into the frequency domain by applying fast Fourier transform (FFT). The transmission loss on the specified channel was evaluated by calculating the attenuation of field strength at the frequency of interest. Table 3.5 shows comparisons between the measured path loss in the monopole-loop case, and link loss results obtained from conformal FDTD modeling for specific on-trunk distances. The path loss was found to be 43.2 dB. This value is comparable to the measurement value (approximately 44 dB, as shown in Figure 3.4) obtained for the monopole-loop case for the trunk-to-trunk scenario when the body is standing upright. Excellent agreement between measurement and simulation was found, on the order from 1 to 2 dB, with a mean discrepancy of 0.2 dB. The small discrepancies can be accounted for by the fact that only a simple monopole model was

Table 3.5 Comparison Between Modeled and Measured On-Trunk Path Loss

Tx/Rx Distance (cm)	Path Loss (dB)	
	Measured (Monopole-Loop)	*Simulated*
20	40.0	40.6
24	40.5	41.6
28	44.0	43.6
31	46.7	44.6

Frequency = 2.45 GHz; body model as in Figure 3.20.

adopted for transmitter, while a probe was actually employed instead of a loop antenna in the measurements. Additionally, the clothing materials and the presence of minor scatterers could lead to the slight changes apparent between measured and modeled results.

Figure 3.19 shows the computed transmission loss between monopole antennas placed on the human arm, when this is considered in isolation, and when modeled as a cone of length 480 mm, with an upper diameter of 114 mm and a lower diameter of 50 mm, and with electrical properties $\varepsilon_r = 50.2$ and $\sigma = 1.6$ S/m. The antennas are placed 280 mm apart and in the same rotational position on the cylinder. It can be seen that there is good agreement between CST software and the Queen Mary FDTD code. In addition, both of these agree well with a measurement using a human arm.

Figure 3.20 presents the electric field distribution around and on the surface of the human body at different time steps, using the Queen Mary FDTD code. Figure 3.20(a) shows the energy surrounding the modeled source (monopole) at early time

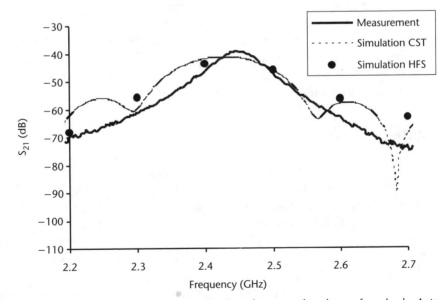

Figure 3.19 Path gain (S_{21}) of on-arm propagation when arm placed away from body. Antennas as in Figure 3.4; frequency = 2.45 GHz; arm modeled as cone: length = 480 mm; diameter at top = 114 mm; diameter at bottom = 50 mm; $\varepsilon_r = 50.2$, $\sigma = 1.65$ S/m.

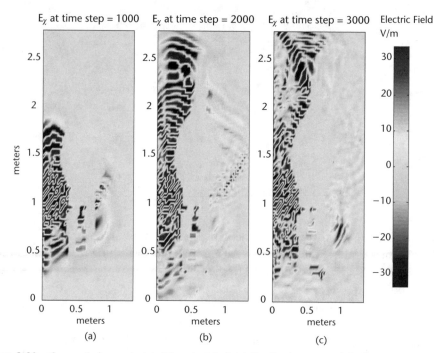

Figure 3.20   Computed snapshots of the electric field distribution around the human body model at different time steps using FDTD: (a) steps = 1,000, (b) steps = 2,000, and (c) steps = 3,000. Frequency = 2.45 GHz; dry skin: ε_r = 38.1, σ = 1.441 S/m; muscle: ε_r = 52.7, σ = 1.705 S/m; lung: ε_r = 34.5, σ = 1.219 S/m. (*From:* [16]. © 2005 ANTEM. Reprinted with permission.)

steps in the computation, with most of the energy concentrated around the source and propagating around the waist in both directions (i.e., following geodesic paths around the body). In Figure 3.20(b), the waves are traveling in both directions along the human body surface and exhibit some attenuation, while the human body shadowing is apparent in Figure 3.20(b, c), with the electric field being stronger at the source side of the body than at the opposite side. This has been verified through on-body measurements with attenuation due to body trunk equal to approximately 19 dB, and attenuation due to the head equal to approximately 13 dB.

3.5   Systems Modeling

Prior to designing a suitable and efficient radio interface for the wireless body area network, it is necessary to assess the performance of existing standards, in order to emphasize the main areas in which new techniques would provide a substantial improvement in a body-wide networked system throughput and reliability. Applying Bluetooth specifications [24], system-level modeling of suitable transmitter and receiver architecture has been performed using measured data similar to that shown in Figure 3.4.

The transceiver schematic of the modeled radio system is shown in Figure 3.21. The system-level modeling architecture of the system was achieved using Agilent RF Designer, and the RF of the system was set to 2.4 GHz. The transmitting and receiving antenna gain was assumed to be 0 dBi. The Bluetooth specification for the modu-

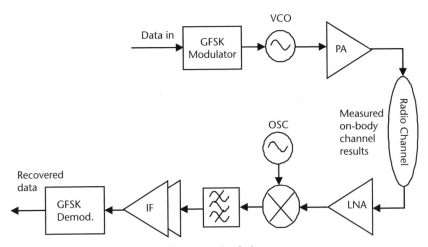

Figure 3.21 Block diagram of Bluetooth system simulation.

lation index requires the application of a BT = 0.5 (bandwidth × period) Gaussian lowpass filter prior to the modulation of the data onto the carrier. For demodulation, the same requirements apply, with the addition of sampling and threshold actions with respect to the whole system delay.

The measured on-body propagation channel results were included in the simulation, and the power applied to the transmit antenna was varied from 0 to −30 dBm, in steps of 10 dB. As mentioned earlier, the channel path gain variation range reaches 36 dB in the trunk-to-head case and approximately 40 dB for trunk-to-hand path, which causes received signal power levels to exceed the dynamic range specification, affecting the performance of the system and the reliability of the data reception. The bit error ratio (BER) of the system is calculated to evaluate overall system performance. Since the GFSK modulation and detection technique is used, the BER calculation is based on the following equation:

$$P(e) = 0.5 \cdot \exp\left(-\frac{E_b}{2n_0}\right) \tag{3.2}$$

where E_b/n_0 is the bit energy-to-noise density, which is simply calculated from the signal-to-noise ratio and data rate.

To evaluate the system with channel characteristics as close as possible to the real environment, the measured path loss variation was included within the radio link component in the modeled architecture. The BER of the received signal also varies, which causes BER sensitivity levels to decrease, adding more restrictions on system parameters. The BER-calculated performance was different for the various transmit powers applied. In both trunk-to-head and trunk-to-hand (worst case) scenarios, the system performed well for transmit power down to −10 dBm, and acceptable performance was obtained for values down to −20 dBm. However, in the −30 dBm transmit power case, high BER bursts resulted in degraded overall system performance, as shown in Figure 3.22. The percentage of time that the bit error rate is more than 0.1% (i.e., the requirement for unacceptable Bluetooth, short-range wireless communication quality), is presented in Table 3.6, for different

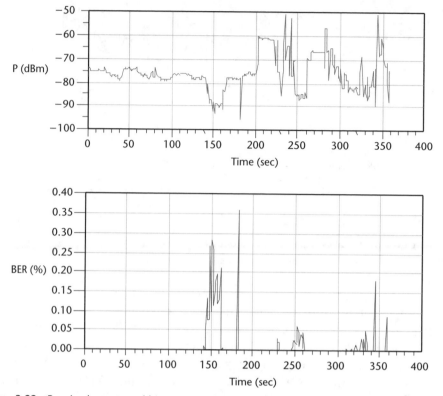

Figure 3.22   Received power and bit error rate for simulation of Bluetooth on body communications link.

Table 3.6   Percentage of Time BER > 0.1% in Simulation of Bluetooth on Body Communications Link

	Trunk-to-Hand	Trunk-to-Head
Transmit Power (dBm)	% Time BER <1e-3	% Time BER <1e-3
0	>99	>99
−10	97	99
−20	89	95
−30	25	77

transmit power values, and for trunk-to-head and trunk-to-hand scenarios. Outage rates are low for powers down to −20 dBm.

3.6   Conclusions

Propagation measurements, models, and simulations relevant to on-body communication links have been presented and discussed. The suitability of a number of antenna types in the context of such links also has been discussed. Simple system modeling has shown the transmit power levels necessary to achieve acceptable bit error rates.

Propagation measurements show that variability in path loss due to different antenna placements and posture changes can be as much as 50 dB. Since the number and relative path gain of communication paths associated with different body postures is highly variable, no single communication channel model is satisfactory. A classification of body postures according to their on-body channel spatial and temporal characteristics is necessary. Due to the limited number of measurements that have been performed, this is indicative of the methodology that needs to be adopted to create such a model.

The dynamic nature and subsequent propagation paths of the body create a significant extra dimension of difficulty, both in modeling and in simulation. At microwave frequencies, using for example FDTD methods, the body is a rather large problem, which is at the limit of what can currently be managed on single PC-based computer. While cluster computing can solve this problem, its extension to modeling the dynamics has not yet been significantly addressed.

The variability of the communication paths means that it is not possible to create a simple specification of the desired antenna characteristics. In this chapter, results for various transmit and receiver antenna combinations over two widely differing body paths have been presented. Antennas have been chosen that represent realizable antenna classes, such as the loop, monopole, patch, and patch array. The results for most of the cases show that the monopole-monopole combination gives the lowest link loss. However, this study is by no means exhaustive. Many more body paths need to be considered, and further optimizations of the antenna may be possible. A low profile disc-loaded monopole antenna has been presented that gives satisfactory results.

Acknowledgments

The authors would like to thank the U.K. Engineering and Physical Research Council for partial support of this project.

References

[1]   King, H. E., and J. L. Wong, "Effects of a Human Body on a Dipole Antenna at 450 and 900 MHz," *IEEE Trans. on Antennas and Propagation*, May 1977, pp. 376–379.

[2]   Chuang, H. R., and W. T. Chen, "Computer Simulation of the Human-Body Effects on a Circular-Loop-Wire Antenna for Radio-Pager Communications at 152, 280, and 400 MHz," *IEEE Trans. on Vehicular Technology*, Vol. 46, No. 3, August 1997, pp. 544–559.

[3]   Salonen, P., Y. Rahmat-Samii, and M. Kivikoski, "Wearable Antennas in the Vicinity of Human Body," *IEEE Antennas and Propagation Society Symp.*, Vol. 1, June 20–25, 2004, pp. 467–470.

[4]   Scanlon, W. G., and N. E. Evans, "Numerical Analysis of Bodyworn UHF Antenna Systems," *Electronics and Communication Engineering J.*, Vol. 13, No. 2, April 2001, pp. 53–64.

[5]   Wang, J., et al., "Computation with a Parallel FDTD System of Human-Body Effect on Electromagnetic Absorption for Portable Telephones," *IEEE Trans. on Microwave Theory and Techniques*, Vol. 52, No. 1, Part 1, January 2004, pp. 53–58.

[6] Iskander, M. E., Z. Yun, and R. Quintero-Illera, "Polarization and Human Body Effects on the Microwave Absorption in a Human Head Exposed to Radiation from Handheld Devices," *IEEE Trans. on Microwave Theory and Techniques*, Vol. 48, No. 11, Part 2, November 2000, pp. 1979–1987.

[7] Bernardi, P., et al., "Specific Absorption Rate and Temperature Elevation in a Subject Exposed in the Far-Field of Radio-Frequency Sources Operating in the 10–900-MHz Range," *IEEE Trans. on Biomedical Engineering*, Vol. 50, No. 3, March 2003, pp. 295–304.

[8] Obayashi, S., and J. Zander, "A Body-Shadowing Model for Indoor Radio Communication Environments," *IEEE Trans. on Antennas and Propagation*, Vol. 46, No. 6, June 1998, pp. 920–927.

[9] Ziri-Castro, K. I., W. G. Scanlon, and N. E. Evans, "Indoor Radio Channel Characterization and Modeling for a 5.2-GHz Bodyworn Receiver," *Antennas and Wireless Propagation Letters*, Vol. 3, No. 1, 2004, pp. 219–222.

[10] Nechayev, Y. I., et al., "On-Body Path Gain Variations with Changing Body Posture and Antenna Position," *IEEE AP-S Int. Symp.*, Washington, D.C., July 2005.

[11] Hall, P. S., M. Ricci, and T. M. Hee, "Measurement of On-Body Propagation Channels," *IEEE Antennas and Propagation Int. Symp.*, June 2002, pp. 310–313.

[12] Hall, P. S., et al., "Characterisation of On-Body Communication Channels," *UNSC/URSI National Science Meeting*, June 2004, p. 155.

[13] Nechayev, Y., et al., "Path Loss Measurements of On-Body Communication Channels," *Int. Symp. Antennas and Propagation*, Sendai, Japan, August 17–21, 2004, pp. 745–748.

[14] Kamarudin, M. R., N. I. Nechayev, and P. S. Hall, "Antennas for On-Body Communications," *IEEE Int. Workshop on Antenna Technology, Small Antennas, Novel Metamaterials*, Singapore, March 2005, pp. 17–20.

[15] Nechayev, Y. I., et al., "Radio Channel Characterisation and Antennas for On-Body Communications," *Loughborough Antenna and Propagation Conference*, Loughborough, U.K., April 2005.

[16] Nechayev, Y. I., et al., "Narrowband and Wideband Radio Channel Characterisation and Antennas for On-Body Communications," *Workshop on Body Sensor Networks*, Imperial College, April 2005.

[17] Nechayev, Y. I., et al., "Antennas and Propagation for On-Body Communication Systems," *11th Int. Symp. Antenna Technology and Applied Electromagnetics—ANTEM*, June 15–17, 2005, Saint Malo, France.

[18] Kamarudin, M. R., Y. I. Nechayev, and P. S. Hall, "Performance of Antennas in the On-Body Environment," *IEEE AP-S Int. Symp.*, Washington, D.C., July 2005.

[19] Zasowski, T., et al., "UWB for Noninvasive Wireless Body Area Networks: Channel Measurements and Results," *IEEE Conference on Ultra Wideband Systems and Technologies*, Reston, VA, 2003.

[20] Ryckaert, J., et al., "Channel Model for Wireless Communication Around Human Body," *Electronics Letters*, April 29, 2004, Vol. 40, No. 9, pp. 543–544.

[21] McNamara, D. A., et al., *Introduction to the Uniform Geometrical Theory of Diffraction*, Norwood, MA: Artech House, 1990.

[22] Hall, P. S., et al., "Microstrip Patch Antenna with Integrated Adaptive Tuning," *10th IEE Antennas and Propagation Conference*, Edinburgh, April 1997, pp. 506–508.

[23] Rostbakken, O., G. S. Hilton, and C. R. Railton, "Adaptive Feedback Frequency Tuning for Microstrip Patch Antennas," *IEE Int. Conference on Antennas and Propagation*, April 4–7, 1995, pp. 166–169.

[24] Bluetooth specifications, http://www.bluetooth.org.

Transmission Mechanism of Wearable Devices Using the Human Body as a Transmission Channel

Koichi Ito, Masaharu Takahashi, and Katsuyuki Fujii

4.1 Introduction to Communications Using Circuits in Direct Contact with the Human Body

In recent years, the development of the information and communication devices, such as cellular phones, personal digital assistants (PDAs), digital video cameras, pocket video games, and so forth, has been progressing rapidly. This evolution gives increased convenience to our daily lives. In the near future, these appliances will begin to be attached to our body, in the form of wearable computers that can be connected to internal and external network systems [1]. However, [2] notes that there is currently no method for these personal devices to directly exchange data. We would like to exchange the data in the wearable devices without any physical constraints, such as external wire connections that may be easily tangled. The solution for networking these personal devices has been proposed as PANs that use the human body as a transmission channel [2, 3]. One of the merits of this system is that data can be exchanged by our natural actions, such as simply touching the receiver, and the user can be clearly aware of connection. Figure 4.1 shows the future applications. We can use these transmission systems for security, electronic money, amusement, and so forth.

Although many studies have been made on the development of wearable devices using the human body as a transmission channel, little is known about the transmission mechanism of such devices in the physical layer, from the viewpoint of the interaction between the electromagnetic wave and the human body [2–14]. Electromagnetic communication trials using the human body as a transmission medium have been carried out for more than a decade. However, most of the research has been conducted by researchers who just want to utilize the results, and until recently the physical mechanisms have not been studied.

This chapter investigates the transmission mechanisms of the wearable devices using the human body as a transmission medium [15–23]. Figure 4.2 shows the communication system [9] of the PAN using a 10-MHz carrier frequency [24]. When a user wearing the transmitter touches the electrode of the receiver, a trans-

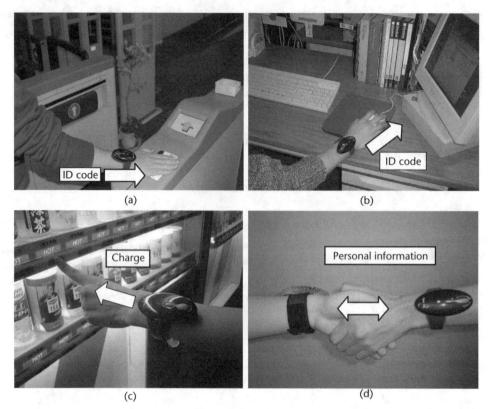

Figure 4.1   Examples of the future applications of PANs: (a) auto lock door, (b) auto login, (c) electronic money, and (d) amusement.

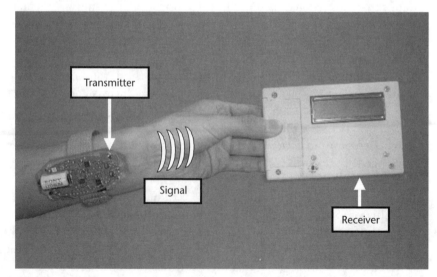

Figure 4.2   Transmission system using the human body as a transmission channel [9].

mission channel using the human body is formed. In this case, the receiver recognizes the user's ID, which can be personalized. This communication system uses the near-field region of the electromagnetic wave generated by the device, which is eventually coupled to the human body by electrodes. The structure of electrodes is one of the key issues for transmission using the human body.

In this chapter, the authors clarify the transmission mechanism of the wearable device using the human body as a transmission channel, from the viewpoint of the interaction between electromagnetic waves and the human body, by using computer simulations and by experiments on a biological tissue-equivalent solid phantom.

In Section 4.2, some calculation models of the transmitter attached to the human body are proposed, using the FDTD method [25, 26] to clarify the transmission mechanism. In the first step, the electric field distributions inside and outside of realistic high-resolution whole-body voxel models of a Japanese adult male and female of average height and weight [27] are investigated. The transmission system described in this chapter uses a 10-MHz carrier frequency, so the wavelength, $\lambda =$ 30m. A whole body calculation area will be needed because the human body may resonate at this frequency. As a result, there is no difference between the electric fields of a male and a female. Moreover, from the viewpoint of computer resources (memory, calculation time, and so forth), the calculation area can be limited to only the arm region, because most of the electric field is concentrated near the tip of the arm.

In the next step, only the arm models are used to simplify the calculation. The calculation models are a male arm, a female arm, and a rectangular parallel-piped homogeneous (muscle) arm that almost imitates the averaged-sized Japanese from finger to elbow [28], respectively. From these results, the authors conclude that the simple homogeneous arm model is sufficient to evaluate the electric field distribution of the human body, although it does not include skin, fat, bone, and so forth.

In the third step, the effective electrode structure is proposed to send the signal from the transmitter to the receiver. The differences of the distributions of the electric field are shown from the viewpoints of impedance matching theory by introducing the equivalent circuit models [29].

In Section 4.3, calculation results are compared to the measured results by using a biological tissue-equivalent solid phantom [30] to show the validity of the calculation. In the first step, each direction (rectangular coordinate) of the current distribution inside the arm is investigated, in order to understand the transmission mechanism. However, it is difficult to directly measure the current distribution inside the human body. It is inferred from measuring the magnetic field distribution close to the human body by using Ampère's law. In order to clarify the validity of the measurement, the measured magnetic field strength is compared to the calculated one. The result indicates a good agreement between measurement and calculation.

In the second step, a portable receiver, which was made by K. Fujii under Mr. S. Tajima's (Sony CSL) instructions, is introduced. This receiver can measure the received signal voltage directly without connecting outer measurement equipment, such as an oscilloscope, which would affect the measurement results. By attaching the receiver to the tip of the arm, electric field distributions and received signal levels are investigated. The reason for discussing the electric field distribution is that the received signal voltage of the receiver is calculated from the electric field. The argument from the viewpoint of the electric field is essential. From these results, the effective direction of electrodes of the transmitter to use the human body as a transmission channel is proposed.

In Section 4.4, after the validity of the calculation model was demonstrated in the previous section, the authors clarify the dominant signal transmission channel, because the question of whether the dominant signal channel is inside or outside the arm still remains unsettled. To answer this question, the calculation model of the arm wearing the transmitter and the receiver placed into a hole of a conductor plate is proposed. The electric field distribution and received signal voltage are investigated as a function of the gap between the hole of the conductor plate and the surface of the arm, when the signal passed through the hole in the conductor plate. If the dominant signal channel is outside the arm, then the received signal is not generated when the gap between the conductor plate and surface of the arm does not exist. On the other hand, if the dominant signal channel is inside the arm, then the received signal is generated in the same condition.

In Section 4.5, the authors conclude their studies concerning the transmission mechanism of the wearable devices using the human body as a transmission channel.

4.2   Numerical Analysis and Equivalent Circuit Models

4.2.1   Whole Body Models

Studies of wearable computers have recently attracted much public attention. It is thought that computing in the near future will be mainly performed through the interaction between wearable computers and ubiquitous computers. The communication system that uses the human body as a transmission channel has been proposed as one of these studies. When a user wearing the transmitter shown in Figure 4.3 touches the electrode of the receiver, a transmission channel is formed via the human body. The transmitter has two electrodes. One is the signal electrode fed by an excitation signal ($3V_{p-p}$, 10 MHz, sinusoidal wave), and the other is the ground (GND) electrode that is connected to the ground level of the electric circuit.

Figure 4.4 shows the calculation model of the transmitter. The authors have focused on the modeling of the transmitter for the FDTD calculations, and there has been considerable validity in that result [16, 20]. Two electrodes and circuit boards are modeled as perfect electric conductors. A continuous sinusoidal wave ($3V_{p-p}$, 10 MHz) is fed to the signal electrode.

In our study, the transmitter uses a 10-MHz carrier frequency, and its wavelength λ is much longer than the height of the human body. In order to investigate the coupling between the transmitter and the human body, full-scale human models are utilized. Figure 4.5 shows the realistic high-resolution whole-body voxel models of a Japanese adult male and female of average height and weight. According to [27], the male is 22 years old, is 172.8 cm tall, and weighs 65.0 kg. The female is 22 years old, is 160.0 cm tall, and weighs 53.0 kg. The resolution of these calculation models is 2 × 2 × 2 mm. By putting these calculation models into the FDTD method, the electric field distributions inside and outside the whole bodies are investigated. The relative permittivity and conductivity of each tissue are equal to the value at 10 MHz [31]. The transmitter is attached to the wrist of the left arm. The size of the cells is $\Delta x = \Delta y = \Delta z = 2$ mm. The absorbing boundary condition is

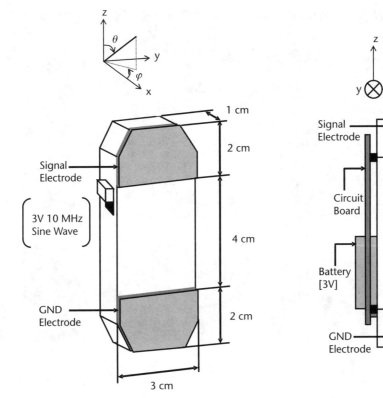

Figure 4.3   Wearable transmitter.

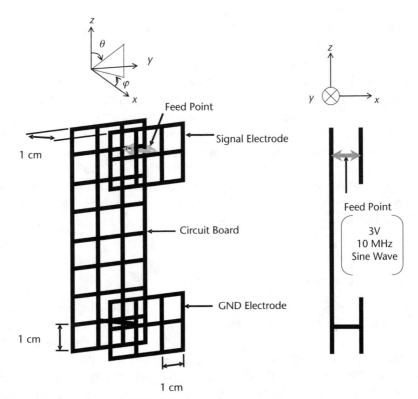

Figure 4.4   FDTD calculation model of the transmitter.

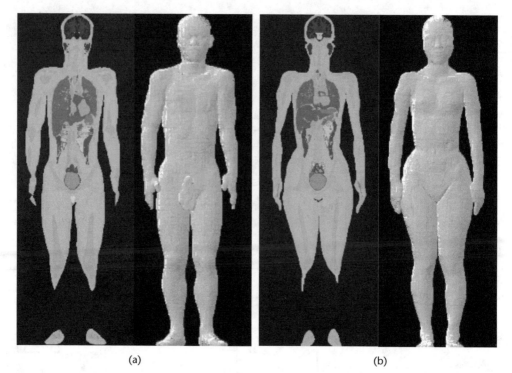

Figure 4.5   Realistic high-resolution whole-body models of a Japanese adult: (a) male, and (b) female [27].

assumed to be Mur's second order, and the time step is 3.84 ps, satisfying the Courant stability condition.

Figure 4.6 shows the electric field distributions inside and outside the whole body of the male and the female. The observation plane includes the feeding point of the transmitter. From these results, it can be seen that most parts of the electric field are concentrated around the tip of the arm. Thus, only the arm model is sufficient to analyze the transmission between transmitter and receiver from the viewpoint of the computer resources (calculation time, memory, and so forth). Moreover, this system has an advantage over those using airborne radio waves from the viewpoint of energy consumption. As a result, this communication system that uses the human body as a transmission channel is useful for personal area networks. In the next section, the differences in the transmission properties caused by the electrode structure will be considered in detail by using only the arm model.

4.2.2   Arm Models

In this section, only the arm model is used, which simplifies the calculation. The electric field distributions inside and outside the arms are calculated. The calculation models are a male arm, a female arm, and a rectangular parallel-piped homogeneous (muscle) arm that almost imitates the averaged-sized Japanese arm from finger to elbow [28]. The size of the cells is $\Delta x = \Delta y = \Delta z = 2$ mm. The absorbing boundary condition is assumed to be Mur's second order, and the time step is 3.84 ps, satisfy-

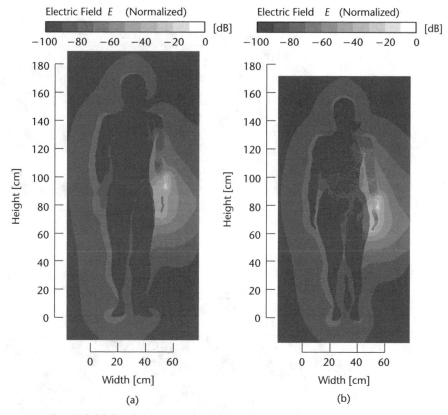

Figure 4.6   Electric field distributions inside and outside the whole body: (a) male, and (b) female.

ing the Courant stability condition. Figure 4.7(a) shows the electric field distribution of the cross section of the male arm. The observation plane includes the feeding point of the transmitter, and the value of the electric field is normalized by the electric field strength at the feeding gap. The electric field is propagated along the length of the arm. Most parts of the electric field are concentrated at the position of the transmitter. This transmission system using the human body as a transmission channel has the advantage of sending the signal merely by touching the electrode of the receiver. It is also advantageous from the viewpoint of the security, because the signal is not radiated into the air but propagated along the length of the arm. Figure 4.7(b) shows the electric field distribution of the cross section of the female arm. The electric field distribution is almost the same as Figure 4.7(a), although the shape and thickness of the arm are different. Figure 4.7(c) shows the simplest calculation model, which is a rectangular parallel-piped homogeneous (muscle) arm. The electric field distributions of Figure 4.7(c) inside and outside the arm are almost the same as those of Figure 4.7(a, b), in spite of a homogeneous structure, not including items such as skin, fat, bone, and so forth, which have different electrical parameters. From these results, it can be concluded that the simplest calculation model in Figure 4.7(c) is sufficient to evaluate the electric field distribution of the human body. After this section, the rectangular parallel-piped homogeneous (muscle) arm is utilized to simplify the discussion on transmission properties.

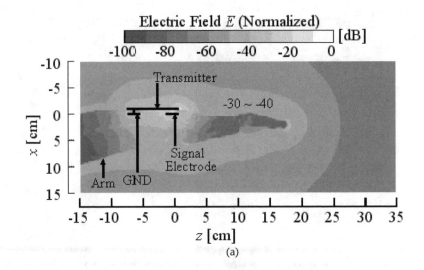

(a)

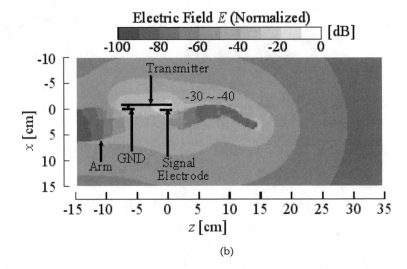

(b)

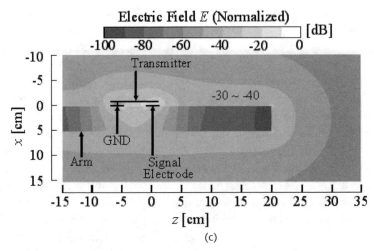

(c)

Figure 4.7 Electric field distributions inside and outside the arm: (a) male, (b) female, and (c) homogeneous model (muscle).

4.2.3 Effective Electrode Structure

This section examines the effective electrode structure of the transmitter for using the human body as a transmission channel. To investigate the electric field difference due to the electrode structure, two types of the electrode structures are introduced. Figure 4.8 shows the calculation model of the arm with the transmitter. Figure 4.8(a) shows the electrode model with the GND electrode, and Figure 4.8(b) shows the electrode model without the GND electrode. The arm is modeled as a rectangular parallel-pipe ($5 \times 5 \times 45$ cm), and the electrical parameters are equal to those of the muscle (relative permittivity $\varepsilon_r = 170.73$, and conductivity $\sigma = 0.62$ S/m) [32]. The size of the cells is $\Delta x = \Delta y = \Delta z = 1$ cm. The absorbing boundary condition is assumed to be Mur's second order, and the time step is 19.2 ps, satisfying the Courant stability condition.

Figure 4.9 shows the result of the electric field distributions inside and outside the arm. The observation plane is the x-z plane at $y = 0$, and 0 dB indicates the strength of the electric field at the feeding point. Figure 4.9(a) shows that the electric field is distributed along the surface of the arm (approximately from −30 to −40 dB), because the current path is formed between the signal electrode and the GND electrode. However, in Figure 4.9(b), level of the electric field on the surface of the arm

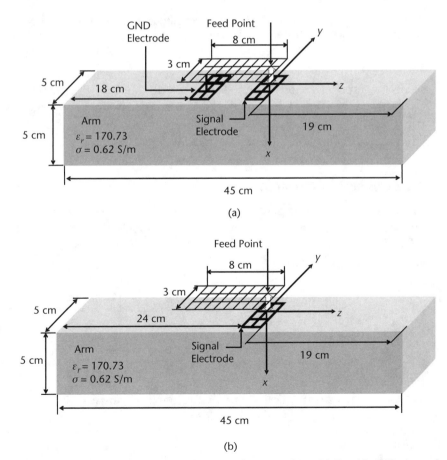

(a)

(b)

Figure 4.8 Calculation model of the arm wearing the transmitter: (a) Tx with GND electrode, and (b) Tx without GND electrode.

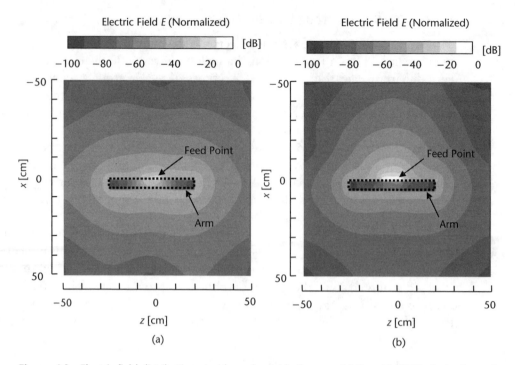

Figure 4.9   Electric field distributions inside and outside the arm: (a) Tx with GND electrode, and (b) Tx without GND electrode.

seems low (approximately from −40 to −50 dB), and the electric field does not penetrate inside the arm. Therefore, the GND electrode is necessary to generate the electric field around the arm. Next, this result will be investigated by using the equivalent circuit models of power transmission.

4.2.4   Equivalent Circuit Models

In this section, the differences of the electric field distributions in Figure 4.9 are investigated by using the equivalent circuit models [29]. Figure 4.10 shows the equivalent circuit models of the transmitter attached to the arm. Input power is expressed as the following equation:

$$P_{in} = \frac{1}{2}\text{Re}\left\{ Z_{in} \cdot \frac{V_g}{Z_g + Z_{in}} \cdot \frac{V_g^*}{\left(Z_g + Z_{in}\right)^*} \right\} = \frac{1}{2}\frac{|V_g|^2 \cdot \text{Re}(Z_{in})}{|Z_g + Z_{in}|^2} = P_{av}S \tag{4.1}$$

where V_g, Z_g, Z_{in}, and P_{av} are the supply voltage, the output impedance, the input impedance, and the available power, respectively. If Z_g is equal to Z_{in}, then Figure 4.10 represents a transmission line that is matched, and $S = 1$. Therefore, P_{av} and S are expressed as the following equations.

$$P_{av} = \frac{|V_g|^2}{8\text{Re}(Z_g)} \tag{4.2}$$

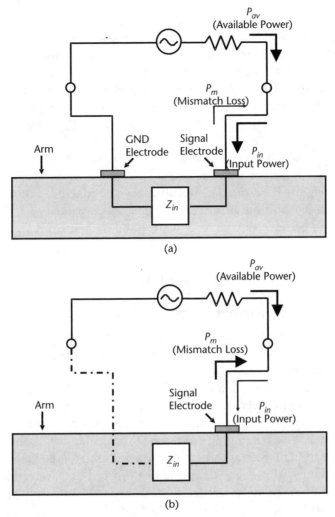

Figure 4.10 Equivalent circuit models: (a) Tx with GND electrode, and (b) Tx without GND electrode.

$$S = \frac{4\mathrm{Re}(Z_g) \cdot \mathrm{Re}(Z_{\mathrm{in}})}{\left|Z_g + Z_{\mathrm{in}}\right|^2} \tag{4.3}$$

Under the assumption that the resistance of the electrode can be neglected, total power of the loss and the radiation efficiency based on the available power are expressed as follows:

$$P_t = P_b + P_m \tag{4.4}$$

$$\eta = \frac{P_r}{P_{av}} = \frac{P_{av} - P_t}{P_{av}} \tag{4.5}$$

where P_b, P_m, and P_r are the absorption power of the human body, the mismatch loss, and the radiation power, respectively.

By using the FDTD calculation, the input impedance Z_{in} [Ω], the input power P_{in} [W], and the radiation power P_r [W] can be calculated. Table 4.1 shows the partition of the available power P_{av}, where $P_{av} = P_{in} + P_m$, under the assumption that the output impedance of the equivalent circuits are 50Ω, by substituting Z_{in}, P_{in}, and P_r for (4.1) to (4.5). From Table 4.1, the imaginary part of Z_{in} is low in the case of the GND electrode. Thus, most parts of the available power P_{av} are fed to the signal electrode, and change the absorption power of the human body P_b. On the other hand, in the case without the GND electrode, the input impedance Z_{in} has a large amount of capacitance, because the absence of the GND electrode causes stray capacitance between the human body and the transmitter. Thus, most of the available power P_{av} is lost as mismatch loss P_m. From these results, it can be concluded that existence of the GND electrode can be quite effective for signal transmission, because it enables impedance matching between the signal generator and the human body.

4.3 Experiments Using Human Phantom

4.3.1 Measurement of the Signal Distributions

A phantom is the material that has the same physical properties as biological tissues and the human body. Accurate measurements for the electromagnetic field distribution in the human body are conducted using various biological tissue-equivalent phantoms, such as liquid or solid phantoms, in which nonuniform models are applicable [33]. In this section, the current distributions inside the human body are measured by using the solid phantom that imitates the human arm introduced in Section 4.2, which shows the validity of the FDTD calculations. From these results, the optimal direction of the transmitter electrodes is used on the human body as a transmission channel. However, it is difficult to directly measure the current distribution inside the human body. Rather, it is inferred from measuring the magnetic field distribution close to the human body, by using Ampère's law. A shielded loop antenna with a diameter of 1 cm is used for the measurement.

Figure 4.11 shows the arrangement used to measure the various components of the magnetic field distribution. The transmitter has two electrodes. One is the signal

Table 4.1 Calculated Parameters of the Equivalent Circuits (10 MHz)

	With GND	Without GND
Z_{in} (Ω)	$37.92 - j4.598$	$13.56 - j2.835$
P_{av} (W)	5.625×10^{-3}	5.625×10^{-3}
S	9.784×10^{-1}	3.373×10^{-4}
P_{in} (W)	5.508×10^{-3}	1.898×10^{-6}
P_m (W)	1.212×10^{-4}	5.623×10^{-3}
P_b (W)	5.499×10^{-3}	1.684×10^{-6}
P_r (W)	4.613×10^{-6}	2.128×10^{-7}
η	8.201×10^{-4}	3.783×10^{-5}

$$(P_{av} = P_{in} + P_m \quad P_{in} = P_b + P_r)$$

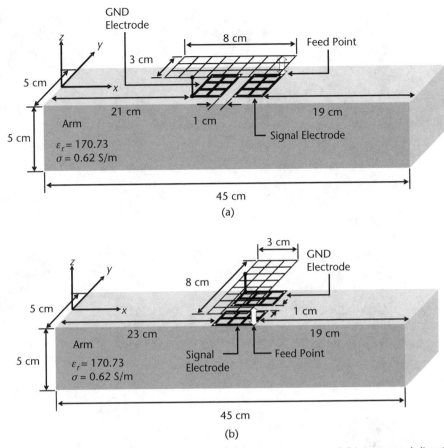

Figure 4.11   Direction of the transmitter: (a) longitudinal direction, and (b) transversal direction.

electrode to feed an excitation signal ($3V_{p-p}$, sinusoidal wave of 10 MHz), and the other is the GND electrode that is connected to the ground level of the electrical circuit. The direction of the transmitter changes according to two patterns to compare the magnetic field distributions. One pattern is the longitudinal direction shown in Figure 4.11(a), and the other pattern is the transversal direction shown in Figure 4.11(b). In addition, the conventional distance between the signal electrode and the GND electrode was 4 cm. However, its distance is reduced to 1 cm, so as to be less than the width of the arm. The experimental muscle-equivalent phantom used for the arm, which is modeled by a rectangular parallel-pipe (5 × 5 × 45 cm), has relative permittivity $\varepsilon_r = 81$ and conductivity $\sigma = 0.62$ S/m. Although the relative permittivity of the muscle at 10 MHz equals 170.73 [32], the authors have verified that they can use this phantom [20], because the received signal voltage is almost the same, and there is great difficulty in making a phantom with such a high relative permittivity. Moreover, in Section 4.2, the electric field distribution around the whole body model wearing the transmitter was investigated, and it was found that the electric field is concentrated around the arm. Hence, an arm model without the whole body can be used.

Figure 4.12 shows the experimental setup for enhanced measurements. For the shielded loop antenna, a coaxial cable and a copper wire with a diameter of 1.6 mm

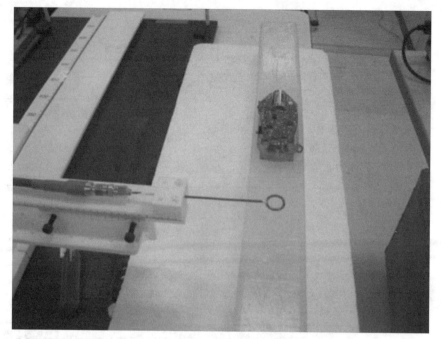

Figure 4.12   Setup for measurement (longitudinal direction).

are used. At the tip of the antenna, a 0.5-mm gap is constructed to generate the received signal voltage. A spectrum analyzer and a power amplifier are also used. The shielded loop antenna is set 3 cm above the surface of the arm, and the magnetic field distributions are measured along the x-$axis$ at $y = 0$ and $z = 3$ cm.

Figure 4.13 shows the various components of the magnetic field distribution around the arm with the transmitter. The level of each point is normalized by the value of H_z at $x = 26$ cm in Figure 4.13(b), which is the maximum value of all measured data in Figure 4.13. In the case of the transmitter set to the longitudinal direction in Figure 4.13(a), from Ampère's law, the dominant current distribution inside the arm is the x component, because the H_y component is stronger compared to the H_x and H_z components near the tip of the arm ($x = 30$ to 45 cm). In the case of the transmitter attached to the transversal direction in Figure 4.13(b), the dominant current distribution inside the arm is the y component, because the H_x and H_z components are stronger than the H_y components near the tip of the arm ($x = 30$ to 45 cm). Therefore, it can be concluded that the direction of the dominant current distribution inside the arm is the same as the direction of the electrodes of the transmitter, because the current is formed between the signal electrode and the GND electrode.

4.3.2   Comparison Between Measurement and Calculation

Figures 4.14 and 4.15 indicate the comparison between measured and simulated magnetic fields to show the validity of the measurements and the FDTD calculations. These figures show the transmitter set to the longitudinal direction and transversal direction, respectively. In the FDTD calculation, two electrodes and circuit boards of the transmitter are modeled as perfect conductor sheets. The sizes of the electrodes are 2 × 3 cm, and the size of the circuit board is 8 × 3 cm. The numer-

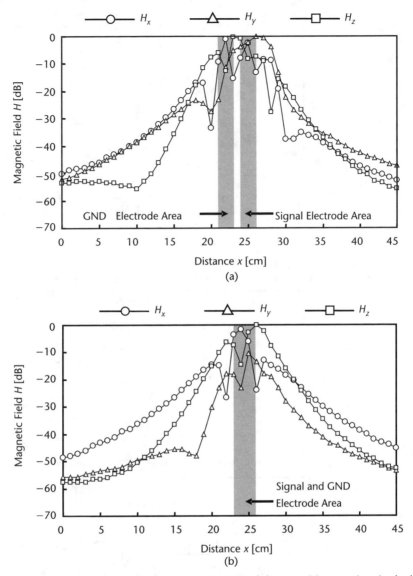

Figure 4.13   Measured magnetic field distributions around the arm: (a) transmitter in the longitudinal direction, and (b) transmitter in the transversal direction.

ical muscle-equivalent phantom used for the arm has relative permittivity $\varepsilon_r =$ 170.73, and conductivity $\sigma = 0.62$ S/m. The size of the cells is $\Delta x = \Delta y = \Delta z = 1$ mm. The absorbing boundary condition is assumed to be Mur's second order, and the time step is 1.92 ps, satisfying the Courant stability condition. In Figures 4.14 and 4.15, all the simulated data are normalized by the value of H_x at $x = 24.5$ cm in Figure 4.15(a), which is the maximum value of all the simulated data in Figures 4.14 and 4.15. To discuss the difference between the measured and calculated magnetic fields, the following equation is defined:

$$\text{Difference} = \sqrt{\frac{1}{x_2 - x_1} \int_{x_1}^{x_2} \left\{ H_{Meas.}(x) - H_{FDTD}(x) \right\}^2 dx} \qquad (4.6)$$

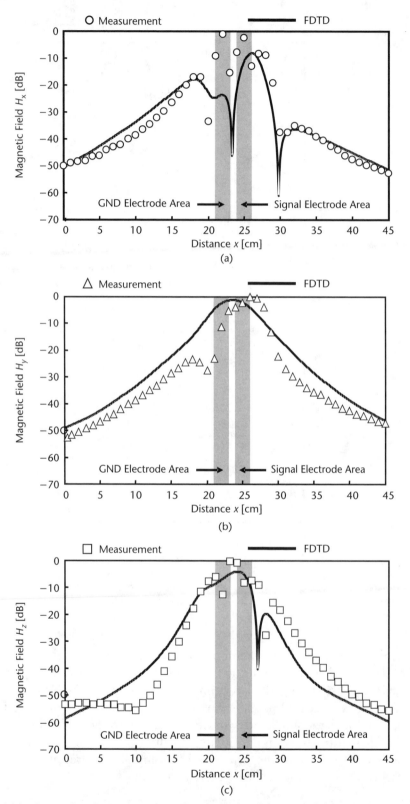

Figure 4.14 Comparison between measured and calculated magnetic field in the longitudinal direction: (a) H_x component, (b) H_y component, and (c) H_z component.

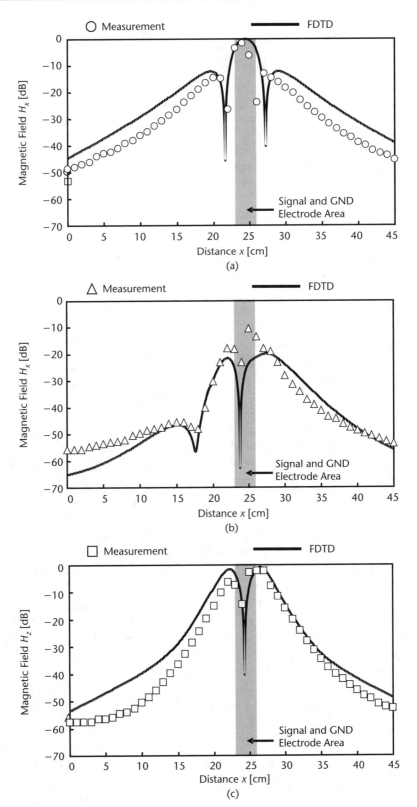

Figure 4.15   Comparison between measured and calculated magnetic field in the transversal direction: (a) H_x component, (b) H_y component, and (c) H_z component.

where $H_{Meas.}(x)$ and $H_{FDTD}(x)$ indicate the measured and calculated magnetic fields, respectively. x_2 and x_1 equal 45 cm and 0 cm, respectively. Table 4.2 shows the difference between $H_{Meas.}(x)$ and $H_{FDTD}(x)$, using (4.6). From Table 4.2, all the differences of Figures 4.14 and 4.15 are almost less than 7 dB. These differences may be caused by the dynamic range for the measurement being limited to approximately −55 dB. However, the rate of decrease and the null point as a function of the distance x is generally in agreement. From this viewpoint, each component of the magnetic field distribution can be compared, and this supports the validity of the measurements and the FDTD calculations.

4.3.3   Electric Field Distributions in and Around the Arm

Figure 4.16 illustrates the electric field distributions (root-sum-square) of both directions of the electrodes of the transmitter. The reason for discussing the electric field distribution is that the received signal voltage of the receiver is calculated from the electric field. Thus, the argument from the viewpoint of the electric field is essential. The structure of the receiver is illustrated in Figure 4.17. The receiver has a receiving electrode and an LCD that can directly indicate the received signal voltage. The reason for having no GND electrode on the receiver is that it reduces the received signal voltage [17]. There is no optimal direction of the receiver because the receiver has only one electrode. Figure 4.17(b) is the FDTD calculation model of the receiver. The receiving electrode and the circuit board are modeled as perfect electric conductors. The received signal voltage is calculated from the electric field at the receiving point. Therefore, this receiver does not detect the magnetic field, but only the electric field. The distance between the transmitter and receiver is fixed at 17 cm, because the transmitter is located at the center of the arm and the receiver is located at the tip of the arm. The observation plane is the x-z plane, including the receiving point of the receiver. The electric field is normalized to the value at the feeding gap. In the case of the longitudinal direction of the transmitter in Figure 4.16(a), the electric field is propagated along the surface of the arm (from −50 to −60 dB). However, in Figure 4.16(b), the level of the electric field on the surface of the arm seems low (from −60 to −70 dB), and the electric field is not propagated along the surface of the arm, but rather is radiated on the upper side of the arm. This result indicates a disadvantage for practical use, compared to Figure 4.16(a), in terms of higher signal reception. In addition, Figure 4.18 shows the electric field distributions without the arm. The loss of the electric field at the receiving point is quite large (< −80 dB) compared to the Figure 4.16. Therefore, it can be concluded that this transmission system uses the human body as a transmission channel.

Next is given an interpretation of the difference between the two parts of Figure 4.16. As shown in Figure 4.13(a), the dominant current distribution is the x compo-

Table 4.2   Difference Between Measured and Calculated Magnetic Field Distributions

	Figure 4.14			Figure 4.15		
	(a)	*(b)*	*(c)*	*(a)*	*(b)*	*(c)*
Difference (dB)	6.7	7.0	7.2	7.0	5.4	6.5

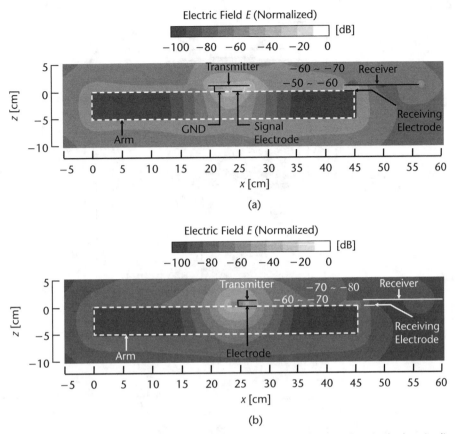

Figure 4.16   Electric field distributions in and around the arm: (a) transmitter in the longitudinal direction, and (b) transmitter in the transversal direction.

nent when the transmitter is set to the longitudinal direction. Thus, the electric field is distributed along the length of the arm (x direction). On the other hand, as shown in Figure 4.13(b), the dominant current distribution is the y component when the transmitter is set to the transversal direction. Thus, the electric field is not distributed along the x direction. The difference of the current distribution causes the difference of the electric field distribution.

4.3.4   Received Signal Voltage of the Receiver

Figure 4.19 shows the measurement arrangements for the received signal voltage according to the direction of the electrodes of the transmitter. In order to verify the validity of the calculation models, the received signal voltage is compared to the measured voltage by using the biological tissue-equivalent solid phantom. In Figure 4.19(a), the transmitter is attached in the longitudinal direction, while in Figure 4.19(b), the transmitter is attached in the transversal direction.

Figure 4.20 shows the comparison between the measured received signal voltages and the calculated voltages. Good agreement is noted, which supports the validity of both the FDTD model and measurement. When compared to the received signal voltage for the longitudinal direction, the transversal voltage drops by nearly

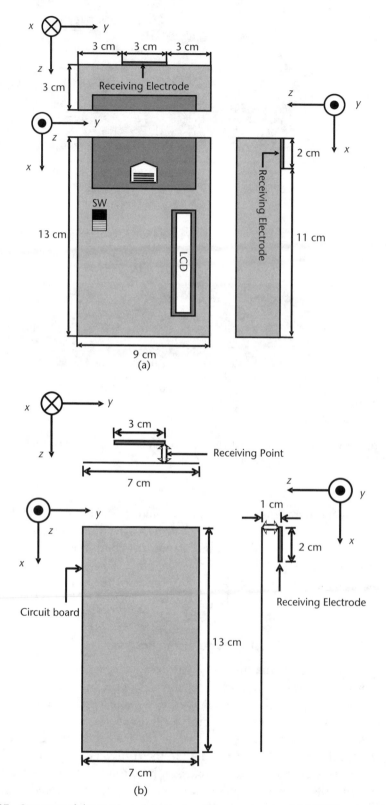

Figure 4.17 Structure of the receiver: (a) exterior of the receiver, and (b) FDTD calculation model.

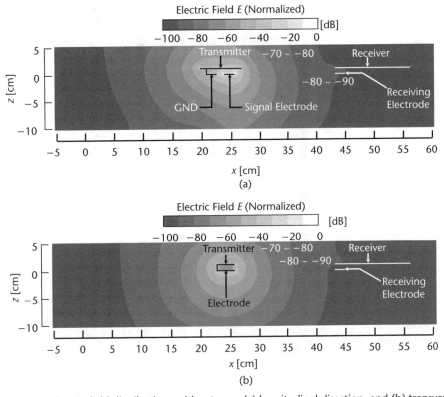

Figure 4.18   Electric field distributions without arm: (a) longitudinal direction, and (b) transversal direction.

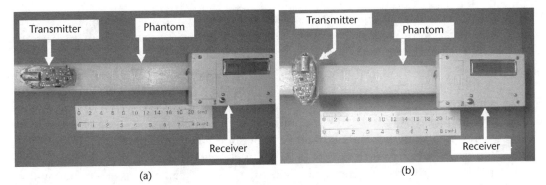

Figure 4.19   Measurement conditions of the received signal voltage: (a) longitudinal direction, and (b) transversal direction.

10%. The received signal voltages without the arm are almost zero. The transmission system using the human body as a transmission channel has an advantage over transmission systems using airborne radio waves. Regarding the difference of the relative permittivity, the received signal voltages are almost equal. It is appropriate to use the phantom with ε_r set to 81 as a substitute for a phantom with ε_r set to 170.73.

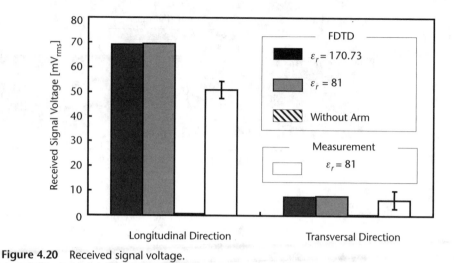

Figure 4.20   Received signal voltage.

In conclusion, the longitudinal direction is more effective for sending the signal to the receiver, as compared with the transversal direction. These investigations have made it clear that we can effectively use the human body as a transmission channel by guiding the current along the length of the arm.

4.4   Investigation of the Dominant Signal Transmission Path

4.4.1   Calculation Model

In this section, the dominant signal transmission channel is investigated, because the question of whether the dominant signal channel is inside or outside the arm still remains unsettled. To answer this question, the calculation model of an arm wearing the transmitter and the receiver placed into a hole of a conductor plate is proposed. Figure 4.21 shows the calculation model of the arm with the transmitter and receiver using the FDTD method. The reason for constructing such a calculation model is as follows. If the dominant signal transmission channel is inside the arm, then the received signal will be generated when the gap $g = 0$. On the other hand, if the dominant signal transmission channel is outside the arm, the electric field from the transmitter will not propagate toward the receiver, but will be reflected at the position of the conductor plate when the gap $g = 0$. The dominant signal transmission path can be clarified by using this calculation model. The size of the conductor plate d is physically infinity, because it is attached to the absorbing boundary of the FDTD. The size of the cells is $\Delta x = \Delta y = \Delta z = 1$ mm. The absorbing boundary condition is assumed to be Mur's second order, and the time step is 1.92 ps, satisfying the Courant stability condition. The distance between the signal electrode and the GND electrode is set to the conventional size (4 cm). By using this model, the electric field distribution and the received signal voltage are investigated as a function of the gap g between the hole of the conductor plate and the surface of the arm.

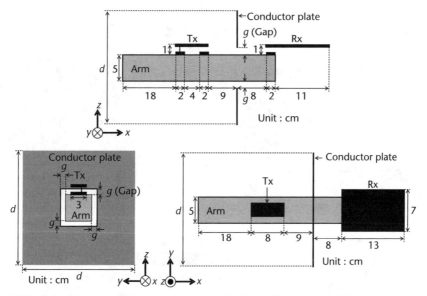

Figure 4.21   Calculation model to investigate the dominant signal transmission path.

4.4.2   Electric Field Distributions and Received Signal Voltages

Figure 4.22 illustrates the electric field distributions (root-sum-square) inside and outside the arm. The observation plane is the x-z plane, and the y plane includes the receiving point. The electric field is normalized to the value at the feeding gap. Figure 4.22(a, b) shows the electric field distributions when the gap g is -1 and 0 mm, respectively. It can be seen that the electric field is not propagated toward the receiver, but instead is reflected at the point of the conductor plate. However, in Figure 4.22(c–f), as the gap g between the conductor plate and the surface of the arm becomes wider, the electric field is propagated more toward the receiver.

Figure 4.23 shows the comparison between the measured received signal voltages and the calculated values as a function of the gap g. To measure the received signal voltages, the conductor plate with a size of 200×200 cm is used ($d = 200$ cm), as shown in Figure 4.24. When the size of the gap g ranges from -1 to 0 mm, the received signal voltage is almost zero. However, as the gap g between the conductor plate and the surface of the arm becomes wider, the received signal voltage rises sharply. The result shows a good agreement between the calculated and measured received signal levels, which indicates a considerable validity in both the FDTD and measurement. The reason for the difference of the received signal voltage between the longitudinal direction in Figure 4.20 and without the conductor in Figure 4.23 comes from the difference of the distance between the signal electrode and the GND electrode. In Figure 4.20, the distance between the signal electrode and the GND electrode is 1 cm. The circuit of the transmitter is almost shorted. Thus, the electric field generated from the transmitter is lower than the conventional size (4 cm).

On the basis of these results, the dominant signal transmission channel using the human body as a transmission channel is not inside the arm, but is on the surface of the arm, because the signal seems to be distributed as a surface wave.

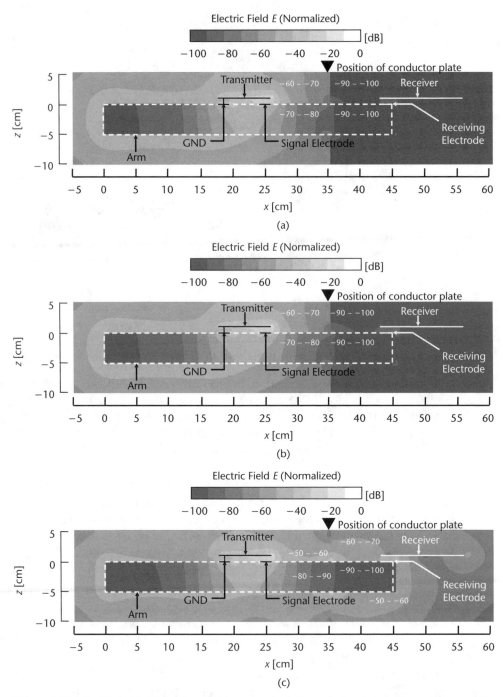

Figure 4.22 Electric field distributions inside and outside the arm wearing the transmitter and the receiver. (a) Gap $g = -1$ mm; (b) gap $g = 0$ mm; (c) gap $g = 1$ mm; (d) gap $g = 5$ mm; (e) gap $g = 10$ mm; and (f) without conductor plate.

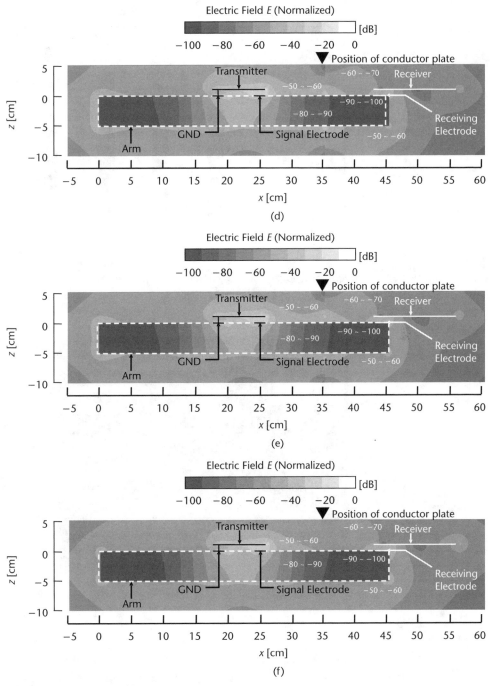

Figure 4.22   (continued.)

4.5   Conclusions

In this chapter, the authors have clarified the transmission mechanism of the wearable device using the human body as a transmission channel, from the viewpoint of the interaction between electromagnetic waves and the human body.

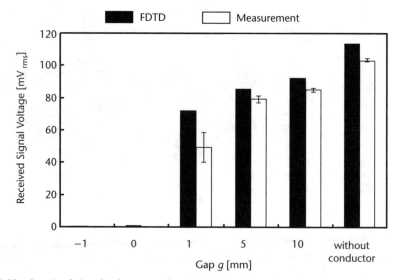

Figure 4.23   Received signal voltage as a function of the gap g between the conductor plate and surface of the arm wearing the transmitter and the receiver.

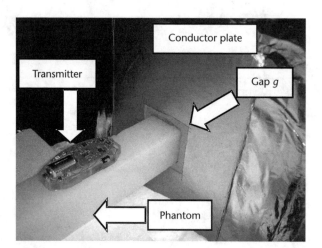

Figure 4.24   Measurement condition.

In Section 4.2, some calculation models and equivalent circuit models of the transmitter attached to the human body were proposed using the FDTD method. The difference in the electric field distributions due to the electrode structures of the transmitter was estimated. As a result, it was found that the GND electrode of the transmitter attached to the arm strengthens the generated electric field around the arm, because it enables impedance matching between the transmitter and the human body.

In Section 4.3, calculation results are compared to the measured results by using the biological tissue-equivalent solid phantom to show the validity of the calculation. Through the measurements, the distributions of the current flowing inside the arm were inferred. The results showed a good agreement between the calculations and the measurements. Setting the two electrodes of the transmitter in the longitudinal direction of the arm is more effective than setting the electrodes in the transversal direction. These investigations have made it clear that we can effectively use the human body as a transmission channel by guiding the current along the length of the arm.

In Section 4.4, the dominant signal transmission channel was investigated, because the question of whether the dominant signal channel was inside or outside the arm still remained unsettled. The results lead us to the conclusion that the dominant signal transmission channel of a wearable device using the human body as a transmission channel is near the surface of the arm, because the signal seems to be propagated as a surface wave. However, there is still been room for theoretical arguments and further study about surface propagation of this result.

References

[1]   Weiser, M., "The Computer for the Twenty-First Century," *Scientific American*, September 1991, pp. 66–75.

[2]   Zimmerman, T. G., "Personal Area Networks (PAN): Near-Field Intra-Body Communication," M.S. thesis, Cambridge, MA: MIT Media Laboratory, September 1995.

[3]   Zimmerman, T. G., "Personal Area Networks: Near-Field Intra-Body Communication," *IBM Systems J.*, Vol. 35, No. 3 & 4, 1996, pp. 609–617.

[4]   IBM Web site, http://www.almaden.ibm.com/cs/user/pan/pan.html.

[5]   Zimmerman, T. G., et al., "Applying Electric Field Sensing to Human-Computer Interfaces," *Proc. CHI'95*, May 1995, pp. 280–287.

[6]   Post, E. R., et al., "Intrabody Buses for Data and Power," *Proc. ISWC'97*, October 1997, pp. 52–55.

[7]   Fukumoto, M., and Y. Tonomura, "Body Coupled FingeRing: Wireless Wearable Keyboard," *Proc. ACM CHI'97*, March 1997, pp. 147–154.

[8]   Handa, T., et al., "A Very Low-Power Consumption Wireless ECG Monitoring System Using Body as a Signal Transmission Medium," *Proc. 1997 Int. Conference on Solid-State Sensors and Actuators*, Chicago, IL, 1997, pp. 1003–1006.

[9]   Matsushita, N., et al., "Wearable Key: Device for Personalizing Nearby Environment," *Proc. 4th Int. Symp. Wearable Computers (ISWC 2000)*, October 2000, pp. 119–126.

[10] Doi, K., et al., "Development of the Communication Module Used Human Body as the Transmission Line," *Proc. Human Interface Symp. 2001*, 2001, pp. 389–392 (in Japanese).

[11] Hachisuka, K., et al., "Development of Wearable Intra-Body Communication Devices," *Sensors and Actuators A: Physical*, Vol. 105, No. 1, 2003, pp. 109–115.

[12] Shinagawa, M., et al., "A Near-Field-Sensing Transceiver for Intra-Body Communication Based on the Electro-Optic Effect," *Proc. Instrumentation and Measurement Technology Conference*, May 2003, pp. 296–301.

[13] NTT redtacton Web site, http://www.redtacton.com/.

[14] Matsushita Electric Works Technology: press release (2004.09.13), http://www.mew.co.jp/e-press/2004/0409-02.htm.

[15] Fujii, K., K. Ito, and S. Tajima, "Signal Propagation of Wearable Computer Using Human Body as Transmission Channel," *Proc. 2002 Interim Int. Symp. Antennas and Propagation*, Yokosuka Research Park, Japan, November 2002, pp. 512–515.

[16] Fujii, K., K. Ito, and S. Tajima, "A Study on the Modeling of Communication System Using Human Body as Transmission Channel," *J. of the Institute of Image Information and Television Engineers (ITE)*, Vol. 56, No. 11, November 2002, pp. 1845–1849 (in Japanese).

[17] Fujii, K., K. Ito, and S. Tajima, "A Study on the Receiving Signal Level in Relation with the Location of Electrodes for Wearable Devices Using Human Body as a Transmission Channel," *Proc. 2003 IEEE Int. Symp. Antennas and Propagation and USNC/CNC/URSI North American Radio Science Meeting*, Columbus, OH, June 2003, pp. 1071–1074.

[18] Fujii, K., and K. Ito, "Evaluation of the Received Signal Level in Relation to the Size and Carrier Frequencies of the Wearable Device Using Human Body as a Transmission Channel," *Proc. 2004 IEEE AP-S Int. Symp. Antennas and Propagation and USNC/URSI National Radio Science Meeting*, Vol. 1, Monterey, CA, June 2004, pp. 105–108.

[19] Fujii, K., et al., "Study on the Optimal Direction of Electrodes of a Wearable Device Using the Human Body as a Transmission Channel," *Proc. 2004 Int. Symp. Antennas and Propagation*, Vol. 2, Sendai International Center, Japan, August 2004, pp. 1005–1008.

[20] Fujii, K., K. Ito, and S. Tajima, "A Study on the Calculation Model for Signal Distribution of Wearable Devices Using Human Body as a Transmission Channel," *IEICE Trans. on Commun.*, Vol. J87-B, No. 9, September 2004, pp. 1383–1390 (in Japanese).

[21] Fujii, K., et al., "A Study on the Relation Between Surface Wave Component and Received Signal Level of the Wearable Device Using the Human Body as a Transmission Channel," *Proc. 2004 Korea-Japan Joint Conference on AP/EMC/EMT*, Seoul, Korea, November 2004, pp. 271–274.

[22] Fujii, K., et al., "Study on the Transmission Mechanism for Wearable Device Using the Human Body as a Transmission Channel," *IEICE Trans. on Commun.*, Vol. E88-B, No. 6, June 2005, pp. 2401–2410.

[23] Fujii, K., et al., "A Study on the Frequency Characteristic of a Transmission Channel Using Human Body for the Wearable Devices," *Proc. 2005 Int. Symp. Antennas and Propagation*, Vol. 2, Seoul, Korea, August 2005, pp. 359–362.

[24] Sony Computer Science Laboratories, Inc. Web site: http://www.csl.sony.co.jp/index.shtml.

[25] Yee, K. S., "Numerical Solution of Initial Boundary Value Problems Involving Maxwell's Equations in Isotropic Media," *IEEE Trans. on Antennas and Propagation*, Vol. AP-14, No. 3, 1966, pp. 302–307.

[26] Uno, T., *Finite Difference Time Domain Method for Electromagnetic Field and Antenna Analyses*, Tokyo: Corona Publishing, 1998 (in Japanese).

[27] Nagaoka, T., et al., "Development of Realistic High-Resolution Whole-Body Voxel Models of Japanese Adult Male and Female of Average Height and Weight, and Application of Models to Radio-Frequency Electromagnetic-Field Dosimetry," *Physics in Medicine and Biology*, Vol. 49, 2004, pp. 1–15.

[28] "Human Body Dimensions Data for Ergonomic Design," *Report of National Institute of Bioscience and Human Technology*, Vol. 2, No. 1, 1994.

[29] Ogawa, K., Y. Koyanagi, and K. Ito, "An Analysis of the Effective Radiation Efficiency of the Normal Mode Helical Antenna Close to the Human Abdomen at 150 MHz and Consideration of Efficiency Improvement," *Electronics and Communications in Japan, Part 1*, Vol. 85, No. 8, 2002, pp. 23–33.

[30] Matsuda, T., *Hyperthermia Manual*, Tokyo: Magbross Press, 1991, pp. 155–156 (in Japanese).

[31] FCC Web site, http://www.fcc.gov/fcc-bin/dielec.sh.

[32] IFAC Web site, http://niremf.ifac.cnr.it/tissprop/.

[33] Okano, Y., et al., "The SAR Evaluation Method by a Combination of Thermographic Experiments and Biological Tissue-Equivalent Phantoms," *IEEE Trans. on Microwave Theory and Techniques*, Vol. 48, No. 11, November 2000, pp. 2094–2103.

CHAPTER 5

Body-Centric UWB Communications

István Kovács, Yang Hao, Andrew Fort, Maciej Klemm, Gert Pedersen, Patrick Eggers, Philippe De Doncker, and Peter S. Hall

5.1 Overview

The narrowband and wideband propagation mechanisms in body-centric scenarios have been discussed in Chapter 3. UWB is one of the key emerging technologies targeted particularly for short-range and low-power radio communication systems with devices having a small-to-medium form factor. Several proposals exist for UWB air interface solutions, and performance investigations have been carried out in PAN and BAN scenarios [1–3]. The design of these UWB transceivers is strongly dependent on the radio environment(s) in which the communication system is envisioned to operate.

In this chapter, we present novel aspects and research results in the field of simulation, measurement, and modeling of the UWB antennas and radio channels in body-centric user scenarios. The major aspects, which differentiate these body-centric investigations from the traditional UWB radio channel studies, are the user-proximity and the user dynamics typical for handheld/device handling scenarios. These scenarios include the communication links between several small-size devices and/or between small-size devices and a physically larger central unit (e.g., portable base station). The physical size of the devices determines not only the possible dynamics, but also the practical size of the antennas, both having significant influence on the radio propagation channel.

Antennas are an inherent part of any wireless communication system, and for UWB antennas there are several additional requirements that have to be fulfilled for body-centric applications, in terms of both physical size and radiation characteristics in the proximity of the human body. Due to these antenna requirements, the clear distinction between the radio link components (i.e., transmit antenna, and channel and receive antenna) is neither always possible nor always practical. In Section 5.2, we address these issues and present simulation and measurement methodologies for an accurate analysis of small UWB antennas. A novel antenna design is presented that is particularly adapted for body-worn devices.

The body-centric UWB channel investigations aim at disclosing the characteristics of the propagation phenomena in typical user scenarios. Traditional channel measurement and simulation techniques developed for narrow-band and wideband cellular systems cannot be directly used, due to constraints in the electrical size of the devices and the near-field propagation effects inherent in a body-centric system. While numerical simulations allow independent study of the radio link components, in the empirical investigations with real antennas, the separation is not possible anymore. Section 5.3 introduces the methodologies used for static and dynamic channel investigations based on simulations and experimental setups. Typical test user scenarios are proposed in order to better understand the body-centric propagation phenomena.

For physical layer design and optimization purposes, an appropriate radio channel model, including antennas, has to be developed. The body-centric channel models derived, based on extensive antenna and radio channel investigations, are introduced in Section 5.4. The main differences between the body-centric channel models and the classical UWB PAN channel models available in the literature, which do not including the user-proximity effects, are highlighted and discussed.

The UWB body-centric antenna and radio channel aspects described in this chapter are far from being comprehensive. The research results give a first look into the specific issues and possible solutions. Efficient and small-size UWB antenna designs for future sensors and devices are certainly a challenge to be tackled. Further radio channel investigations and parameterizations, including all complex effects of the around-the-body propagation phenomena, are required in order to fully understand the communication medium, and to be able to optimize the next generation WPAN and WBAN radio transceivers.

5.2 Antennas

The majority of the UWB antenna designs for communication systems are all derived from a monopole/dipole, [4–7], are tapered slot antennas [8, 9] or are loop antennas [10]. For practical and low-form factor implementations, the printed planar designs of these antennas are generally proposed. Ceramic chip antenna designs using a dipole configuration also exist. The sectorial loop antenna has been proposed as a compact integration solution [11]. Recently, a purely textile UWB annular slot antenna has been reported [12].

In this section, we focus our attention on the UWB antennas for those body-centric applications in which the design becomes more complicated than for free-space operation scenarios, due to the additional form factor constraints and possibly different radiation characteristics targeted.

5.2.1 Design and Analysis

5.2.1.1 Antenna Parameters

The very wide bandwidth of UWB systems makes the design and evaluation of antennas more difficult than for narrowband systems. Traditional narrowband parameters characterizing antennas, such as return loss, radiation pattern, and polarization, are not directly useful in the characterization of UWB antennas. Therefore, UWB antennas need to be evaluated using other specific parameters, adapted to the large frequency bandwidth and possibly to the targeted UWB system type. The spatial frequency domain transfer function of the antenna is the most commonly used parameter. A UWB antenna also can be characterized in the time domain, based on the impulse response. Together with the waveform driving the antenna, the frequency transfer function also allows one to quantify radiation characteristics of a given UWB antenna design.

In the following, we introduce a set of parameters that were found to be useful when analyzing the UWB antennas presented later in Section 5.2.1.2. In the design process, the traditional antenna parameters, such as radiation efficiency and impedance matching, should be considered as well.

Frequency Domain Transfer Functions   The antenna transmit and receive transfer functions can be defined as [13]:

$$H_{Tx}(\omega, \theta, \varphi) = \frac{E_{rad}(\omega, \theta, \varphi)}{V_{in}(\omega)} \tag{5.1a}$$

$$H_{Rx}(\omega, \theta, \varphi) = \frac{E_{inc}(\omega, \theta, \varphi)}{V_{rec}(\omega)} \tag{5.1b}$$

$$H_{Tx}(\omega, \theta, \varphi) = j\omega H_{Rx}(\omega, \theta, \varphi) \tag{5.1c}$$

where the expressions for $H_{Tx}(\omega, \theta, \varphi)$ and $H_{Rx}(\omega, \theta, \varphi)$ are the frequency- and angle-dependent transmit and receive transfer functions. They relate the radiated electric field intensity $E_{rad}(\omega, \theta, \varphi)$ (for a given polarization) to the antenna driving waveform $V_{in}(\omega)$, and the received voltage on the load $V_{rec}(\omega)$ to the incident electric field $E_{inc}(\omega, \theta, \varphi)$, respectively. Both transfer functions include antenna matching.

Signal Fidelity   Based on the frequency or time transfer function, we are able to calculate pulse distortions introduced by an antenna. In time domain formulation, the fidelity between waveforms $x(t)$ and $y(t)$ is generally defined as a normalized correlation coefficient [14]:

$$F = \text{Max}\left(\frac{\int x(t)y(t-\tau)dt}{\sqrt{\int |x(t)|^2 \, dt \int |y(t)^2| dt}} \right) \tag{5.2}$$

The fidelity factor, F, compares only the shapes of the cross-correlated waveforms but not their amplitudes. In practice, this factor is calculated for a given direction in space in order to fully characterize the spatial radiation properties of an antenna. It should be noted that the fidelity depends on the antenna characteristics and on the excitation signal; thus, it is a system-dependent parameter.

Impulse Response and Time Spread   All UWB antennas have a given impulse response that may affect the actual system design, depending on the bandwidth and time domain signal shape. The impulse response of a UWB antenna is also dependent on spatial direction.

For time domain UWB systems, a helpful parameter describing signal distortion introduced by an UWB antenna is based on the time window enclosing a given energy level of the radiated or received pulse. Since we are interested in capturing as much energy as possible, we compare length of time windows, which include 99% of the pulse energy, $W99$. The time spread is a ratio, $E99(\theta, \varphi)$, between the 99% energy time window lengths of the radiated pulse, $W99_{rad}(\theta, \varphi)$, and the antenna input pulse, $W99_{in}(\theta, \varphi)$. The ratio shows how much of the energy of the pulse is spread compared to the input pulse. Based on the $E99(\theta, \varphi)$ parameter, one can calculate the energy spread of the pulse (radiated or received), and use this information in the system design (e.g., to obtain optimal performance with energy detector receivers). In Section 5.4, we present the results from channel measurements also in terms of the 90% energy delay window, $W90$.

$$E99(\theta, \varphi) = \frac{W99_{rad}(\theta, \varphi)}{W99_{in}(\theta, \varphi)} \qquad (5.3)$$

5.2.1.2 UWB Antennas for Body-Centric Applications

The human body is very lossy at microwave frequencies, and there is no direct transmission through the body. The near-field and far-field radiation characteristics of the UWB antennas play an important role in the body-centric signal propagation. Therefore, UWB antennas for body-centric applications need special attention, due to the proximity of the user's body.

For applications where both low-power transmit and receive devices are located on the body or clothing, the antenna near-field characteristics significantly contribute to the propagation phenomena, and the antenna design has to favor the signal propagation along the surface of the body, and minimize the power losses due to absorption in the body. In applications with communication between body-worn sensors and a larger unit that is not necessarily always located on the body (e.g., PDA, mobile phone, or PC-laptop), the far-field antenna radiation characteristics are also important, and a certain directionality is desired for the body-worn antennas.

It is important to note that in both practical application cases described above, the antennas are placed in the far-field distance. Thus, there is no need to analyze the near-field propagation channel, which is the typical case for near-field communication systems.

The interaction between the human body and an antenna is a complicated problem, and needs to be investigated by means of full-wave numerical electromagnetic solvers. Limited evidence investigating the impact of the human body on UWB antenna characteristics exists in the literature [15–18]. The first UWB body-mounted antenna design with highly decreased sensitivity to the presence of the human body was presented in [19].

In another study, the horn-shaped self-complementary antenna (HSCA) [20] and the planar inverted cone antenna (PICA) [21] were applied in an extensive measurement campaign, which provided a clear picture of the influence of antenna characteristics on propagation channel behavior [7]. Different antenna types are expected to have dissimilar effects on the radio channels, especially in the body area network where the transmitter and the receiver are in close proximity. Antenna near-field characterization is also needed [18].

We present a few representative UWB antenna designs with small form factors, used in recent body-centric radio propagation investigations. Monopole and dipole antennas are easy to model and analyze; thus, they are used as reference antennas. The commercially available UWB antennas suitable for body-centric radio investigations are in essence low-profile monopoles [22], and are not addressed in this section.

Planar UWB Disc Monopole Antenna   The antenna geometry is a planar disc monopole, as shown in Figure 5.1. This planar UWB monopole antenna was designed with the criteria of an impedance bandwidth from 3 to 6 GHz, with a return loss below −10 dB. It is easy to design a UWB monopole antenna when considering only the impedance bandwidth. To achieve the same radiation pattern bandwidth is difficult, due to the significant changes in the antenna pattern at higher frequencies [23]. Plots of the return loss and the H_{TX} function (in direction of maximum radiation) of the UWB disc monopole antenna are shown in Figure 5.2.

Slot Antenna   Another possibility of designing an omnidirectional antenna is the slot configuration. In this example, the antenna is a slot cut in the ground plane, fed by two symmetrically placed 100Ω microstrip lines, which are connected in parallel to the 50Ω feed line. This feeding topology provides very wideband matching for

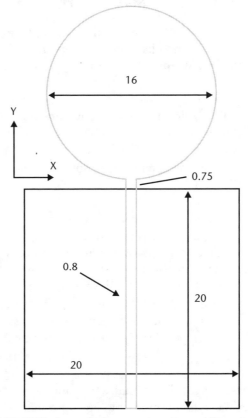

Figure 5.1   Example of the UWB planar disc monopole antenna (dimensions in millimeters). (*From:* [19]. © 2005 IEEE. Reprinted with permission.)

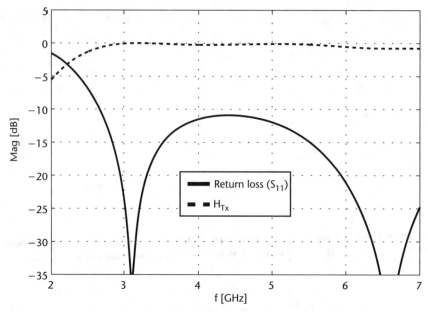

Figure 5.2 Return loss and transmit transfer function for the UWB planar disc monopole antenna in Figure 5.1.

slot and stacked patch antennas [24, 25]. Figure 5.3 shows the slot antenna designed to operate in the lower Federal Communications Commission (FCC) UWB band between 3 and 6 GHz. Due to the small ground plane dimensions compared to the slot dimensions, the antenna has a quasi-omnidirectional radiation pattern in an XZ plane. In Figure 5.4, we present the return loss and transmit frequency transfer function.

A textile UWB annular slot antenna was proposed, which can achieve the same performance as the traditional versions, manufactured using standard PCB technologies [12]. The new feature of this antenna is that it can easily be integrated into clothing, rather than being attached, thus providing the first step towards integration of the electronic elements into clothing, and realization of the so-called system-on-textile (SoT).

Directional Slot Antenna   As mentioned earlier, for certain low-power body-centric applications, it would be desirable to have an antenna with directional far-field radiation characteristics and low near fields radiated towards the body, thus reducing the influence of the body.

In body-centric scenarios, a directive antenna should still be omnidirectional in the plane of the largest dimension along the body surface, and have a moderate front-to-back ratio in the plane of the smallest dimension, perpendicular to the body surface. This configuration does not need any specific directionality of the antennas, and still allows good signal transmission and reception in the directions along the body surface [19].

The geometry of the directional UWB slot antenna is shown in Figure 5.5. The antenna is based on the omnidirectional slot antenna presented in Figure 5.3, and the desired far-field directionality is obtained by adding a reflecting element below the feedline. The small spacing between the reflector and the antenna feed has a

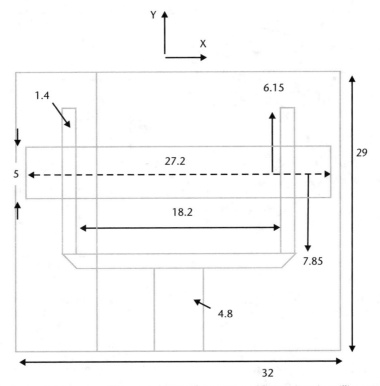

Figure 5.3   Example of an omnidirectional UWB slot antenna (dimensions in millimeters). (*From:* [19]. © 2005 IEEE. Reprinted with permission.)

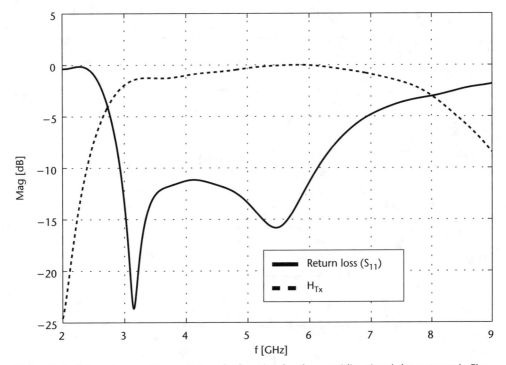

Figure 5.4   Return loss and transmit transfer function for the omnidirectional slot antenna in Figure 5.3.

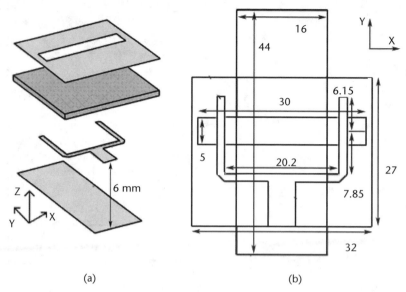

(a)   (b)

Figure 5.5   Low-profile UWB directional slot antenna design (dimensions in millimeters). (*From:* [19]. © 2005 IEEE. Reprinted with permission.)

great impact on the radiation pattern and on the input impedance of the antenna. The physical operation of this directional UWB slot antenna, along with more details about its design and performances, can be found in [19]. The return loss and the transmit transfer function are shown in Figure 5.6.

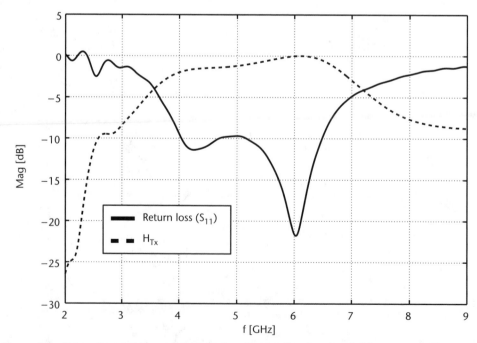

Figure 5.6   Return loss and transmit transfer function for the directional slot antenna in Figure 5.5.

5.2.1.3 Performance Aspects of UWB Antennas for Body-Centric Applications

In Figure 5.7, we present the comparison of the far-field free-space radiation pattern for the UWB antennas presented in Section 5.2.1.1: disc monopole antenna (Figure 5.1), slot monopole antenna (Figure 5.3), and directional slot antenna (Figure 5.5). Compared to the monopole and slot antennas, the UWB directional antenna radiates much more power into the upper hemisphere (i.e., in the direction away from the body).

For all of these three antenna designs, a pulse distortion analysis has been also performed using (5.2) and (5.3). The following quantitative free-space angular characteristics have been analyzed:

- 99% energy window ($E99$), relative to the input pulse;
- 99% energy window ($E99$), relative to the pulse radiated in the direction of maximum gain;
- The fidelity factor of the radiated pulses (F), using the pulse radiated in direction of maximum gain as the reference waveform [19].

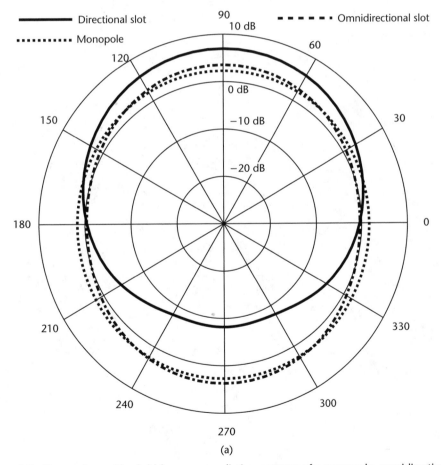

(a)

Figure 5.7 Comparison of far-field free-space radiation patterns of a monopole, omnidirectional slot and directional slot antennas: (a) XZ plane at 3.5 GHz, (b) XZ plane at 4.5 GHz, (c) XZ plane at 6 GHz, (d) YZ plane at 3.5 GHz, (e) YZ plane at 4.5 GHz, and (f) YZ plane at 6 GHz. (*From:* [19]. © 2005 IEEE. Reprinted with permission.)

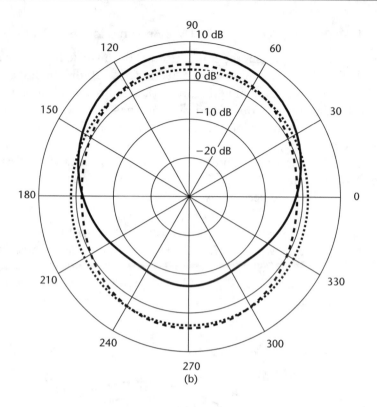

(b)

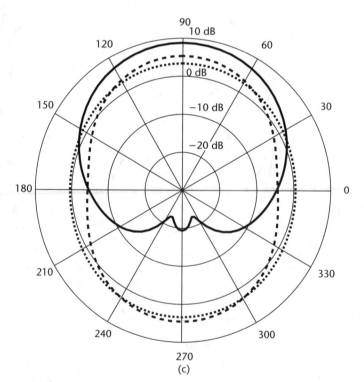

(c)

Figure 5.7 (continued.)

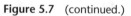

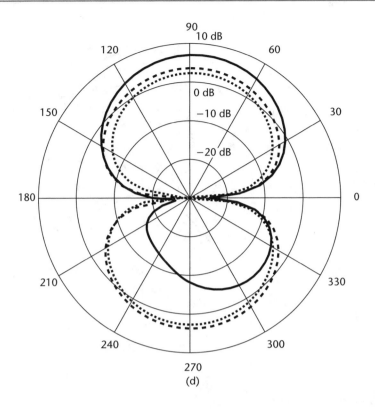

(d)

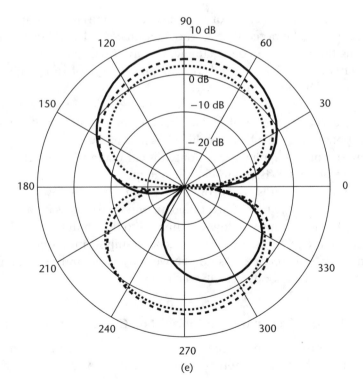

(e)

Figure 5.7   (continued.)

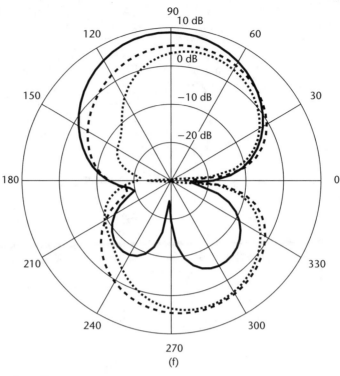

Figure 5.7 (continued).

Representative results of this pulse distortion analysis are presented in Figure 5.8. It should be noted that all these parameters depend on the shape of the pulse driving the antenna. In our example, we have used a modulated Gaussian pulse with main frequency of 5 GHz and effective pulse length of 160 ps, giving a −10-dB pulse bandwidth between 3 to 7 GHz.

We can see that all antennas have fidelity along the XZ plane above the commonly acceptable level of 90%. This indicates that all radiated pulses for spatial directions along XZ plane are alike. It is not the case for the YZ plane, where fidelity is lower, and not symmetrical along this cut plane. This effect is especially visible in the case of the UWB disc monopole antenna.

Considering pulse distortions based on time windows with 99% of the pulse energy, we can see that a directional antenna introduces higher spatial variations of the 99% energy time windows, compared to the omnidirectional antennas. Angular spread of the 99% energy time windows is important if one is interested in applications where any relative position between transmit and receive antennas is possible, such as in typical body-centric scenarios.

5.2.1.4 Body-Proximity Induced Radiation Characteristics and Performances

When operating in a body-centric communication system, the antenna is mounted in close proximity to the human body. It is known that the body has a significant impact on the antenna characteristics [15, 26–31]. Thus, it is important to include the human body model in the antenna design process. For the antenna performance results, we present two different truncated models of the human body:

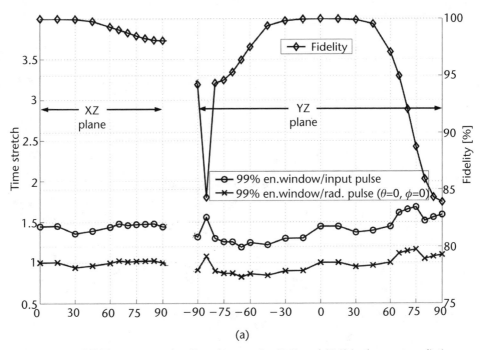

Figure 5.8 Far-field free-space pulse distortion metrics (5.2) and (5.3) in the upper radiation hemisphere (away from the body) for different UWB antennas: (a) planar disc monopole (see Figure 5.1), (b) omnidirectional slot (see Figure 5.3), and (c) directional slot (see Figure 5.5). (*From:* [19]. © 2005 IEEE. Reprinted with permission.)

1. A three-tissue model, consisting of layers of skin (1 mm thick), fat (3 mm thick), and muscle tissue (40 mm thick);
2. A homogeneous model, composed of muscle tissue (44 mm thick).

Overall dimensions of both models are the same: $120 \times 110 \times 44 \text{ mm}^3$. This size for the truncated body model was found by comparing the simulation results in terms of antenna radiation characteristics [19].

To show the effect of the human body on the antenna characteristics, we have simulated the planar disc monopole antenna (Figure 5.1), the omnidirectional slot antenna (Figure 5.3), and the directional slot antenna (Figure 5.5), as presented in Section 5.2.1.2, with the addition of the truncated human body models presented above. For all antennas, we assumed the same distance of 7 mm between the body surface and the antenna feed line. This separation distance, for frequencies between 3.5 and 6 GHz, lies well within the reactive near-field region [32] for all antennas under consideration. It is important to note that for the directional UWB slot antenna, the reflector lies 6 mm below the feed line, making the effective separation between the reflector and the body surface only 1mm.

Figure 5.9 shows the return loss and transmit transfer function (direction of maximum gain) characteristics of UWB antennas, when operating in free space and 7 mm away from muscle tissue or a three-layer-tissue body model.

The impedance bandwidth of the directional UWB slot antenna when operating in free space and close to the human body is almost the same. The only noticeable

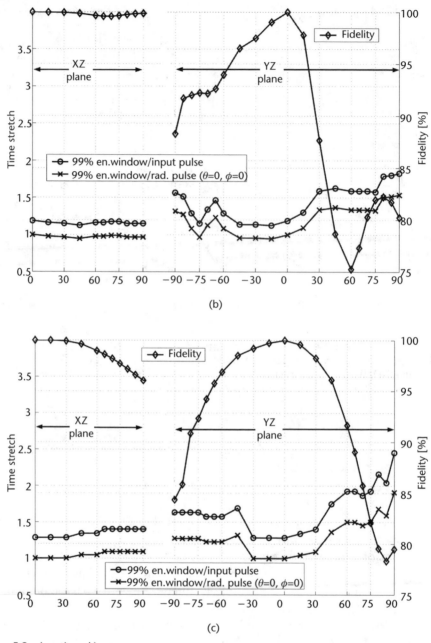

Figure 5.8 (continued.)

difference exists for frequencies below 3.5 GHz. For the omnidirectional UWB slot and monopole antennas, the impedance is greatly changed due to body proximity, and is different with different body models.

Considering transmit transfer functions, we can clearly see that the directional UWB antenna, compared to the omnidirectional antennas (monopole and slot), is much less influenced by proximity to the human body. For the directional UWB antenna, all three transmit transfer functions (i.e., free space, muscle tissue, and 3-layer model cases) are almost identical above 3.5 GHz. Thus, one can expect very

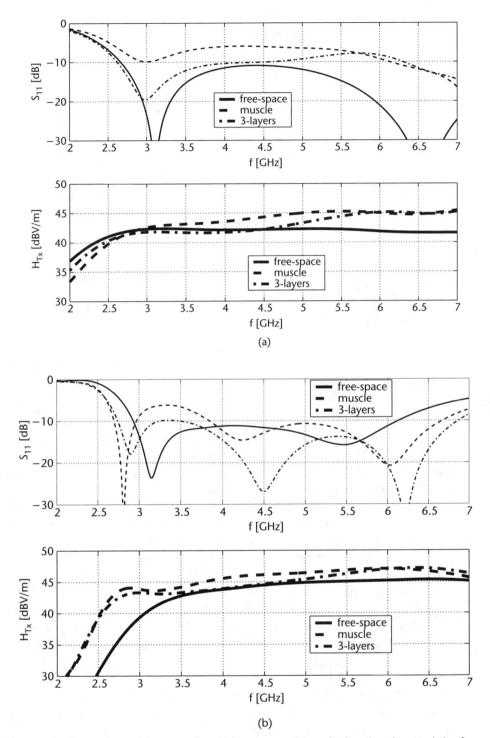

Figure 5.9 Comparison of the return loss (S_{11}) and transmit transfer function characteristics, for UWB antennas operating in free space and in body proximity: (a) planar disc monopole, (b) omnidirectional slot, and (c) directional slot. The transmit transfer function is presented for the direction of maximum gain, and is normalized to the same value for all antennas. (*From:* [19]. © 2005 IEEE. Reprinted with permission.)

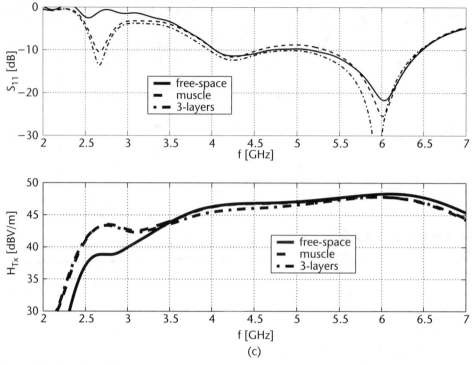

Figure 5.9   (continued.)

similar performances for different on-body positions. For both omnidirectional antennas, higher amplitudes of the transmit transfer function is observed for frequencies above 4 GHz, in spite of low radiation efficiency, due to strong reflection from the human body.

Clearly, the transmit transfer function and return loss results presented in Figure 5.9 can be applied to the receive antenna as well. The observed variations of the transfer functions for the omnidirectional antenna when in proximity to the body are clearly dependent on the body separation distance. When the position of the devices using these antennas varies in time (e.g., through body movements), then the transmitted-received signal will be modulated by the amplitude and phase changes of the transmit and receive transfer functions. This has to be considered in the system design. With the directional antennas, these variations in the transfer functions are greatly minimized, and can be safely disregarded when designing the UWB communication link. From a system design point of view, the signal distortion parameter results in Figure 5.8 are also important.

The selected UWB antenna designs presented also have been characterized in terms of their radiation efficiency, peak SAR values (1g averaged), and representative results, as presented in Table 5.1.

The first main result in Table 5.1 is that antenna performance in terms of the radiation efficiency and SAR has been improved for the directional UWB slot antenna, compared to results for the omnidirectional UWB slot and monopole antenna. Nevertheless, at lower frequencies, there is still a significant amount of power absorbed by the body. Thus, directivity is not necessarily a good solution for body-worn antennas. The reason is that antenna directivity is actually a far-field

Table 5.1 Radiation Efficiency and Peak SAR Values (1g Averaged) of UWB Antennas in the Proximity (7 mm) of a 3-Tissue Body Model [19]; Input Power = 1W

Frequency (GHz)	Monopole		Omnidirectional Slot		Directional Slot	
	Radiation Efficiency (%)	SAR (W/kg)	Radiation Efficiency (%)	SAR (W/kg)	Radiation Efficiency (%)	SAR (W/kg)
3.5	28	38.2	36	30.5	56	13.4
4.5	29	38.6	40	40.7	65	18.3
6	55	30.3	64	31.8	85	11.7

measure, and does not consider the near fields of an antenna. Even if an antenna is directional in one particular angle, there still can be strong near fields in the direction where the human body is located.

The second result in Table 5.1 shows that the radiation efficiency of all antenna types tends to increase with frequency. However, one should be very careful not to treat this statement as a general conclusion for any types/sizes of antennas. When small UWB patch antennas in the proximity (the reactive near-field region) of the same body models were studied [16], results can be different, depending on the antenna size and body separation distance. Moreover, when comparing body models, better efficiencies were generally obtained when the layered model was used.

As a further illustration of the results presented in Table 5.1, we present the power absorbed (in W/m^3) in the skin layer (where the maximum absorption occurs) of the three-layer human body model at 4.5 GHz. Figure 5.10 shows results for both the omnidirectional and directional slot antennas, normalized to the same absolute value. With the use of the reflector element, near fields are scattered, resulting in a lower power deposited in the tissue.

5.2.2 Measurements

The UWB radiation patterns of the antennas discussed in Section 5.2.1 were measured using a 3D spherical antenna measurement system [33, 34], which is very suitable for measurement of a spatial transfer function, as well as for radiation efficiency measurement of free-space and body-worn antennas. Spherical time domain measurements can be performed not only in anechoic rooms, but also in an indoor environment, since the time gating can be used to extract unwanted reflections. A disadvantage, compared to frequency domain measurements, is a reduced dynamic rage of the system [35].

In traditional antenna measurements, the antenna is connected to the measurement equipment with a coaxial cable. This setup is perfectly suited for electrically large antennas. The major problem when measuring electrically small antennas is the current that is induced on the outer surface of the connected cable. This effect causes the cable itself to act as a radiator, significantly changing both the near-field and far-field radiation characteristics of the antenna [36–38].

While the effects of the feed cables can be eliminated or significantly reduced in the case of narrowband antennas [39, 40], it becomes a very difficult task when measuring with small UWB antennas [37]. One possible method of reducing the influence of the cable attached to a UWB antenna is to attach the cable in areas of

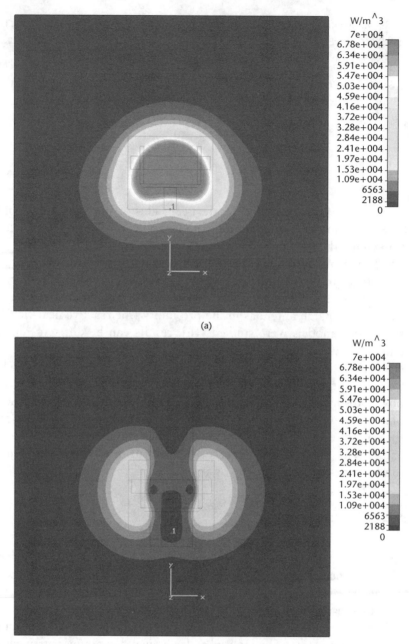

(a)

(b)

Figure 5.10 Power absorbed (W/m³) in the skin layer of the three-layer human body model at 4.5 GHz for: (a) the omnidirectional slot antenna, and (b) the directional slot antenna. Distance between the skin and the feed lines was 7 mm.

low current density [40]. A new method for measurements of small-size wideband and UWB antennas uses an RF-on-fiber optic link (transmitter and receiver), instead of the usual conductive cable feed connections [36, 38, 41]. This setup eliminates the cable connection to the antenna, and can be used for both channel measurements and antenna measurements in the radio anechoic room. However, the proposed

RF-on-fiber setup also adds two new components that have to be mounted on the antenna: the optic receiver and its battery power supply. In the case of the directional slot antenna, full-wave electromagnetic simulations have shown that, in order to minimize the distortions of the radiation patterns, the extra components have to be mounted with their largest dimension perpendicular to the radiating slot, and such that the largest electrical dimension of the original antenna is not extended. If the polarization discrimination of the antenna is also of interest, then a suitable compromise has to be found between the obtainable cross polarization and directivity of the final setup [19].

Given the small electrical size of the antenna, any electrically conductive components (e.g., connectors, cables, integrated components, and so forth), mounted in the real device have strong influence on the antenna radiation characteristics. Thus, analyzing only the performances of the antenna itself or with a different electrical setup can give significantly different results from the real life radiation characteristics in the final application. More details about the integration of the UWB antennas into wearable sensors or hand-held terminals can be found in [42, 43].

5.2.3 Concluding Remarks

In this section, several practical issues have been presented with regards to the analysis, design, and measurement of small form factor UWB antennas for body-centric applications. It was shown that UWB antennas should be evaluated by means of specific parameters, adapted to the large frequency bandwidth, and possibly to the targeted UWB system type, such as frequency domain transfer function, signal fidelity, and impulse response.

Two omnidirectional and one directional UWB antenna designs have been presented: a planar disc monopole antenna, a slot antenna, and a directional slot antenna. The radiation characteristics and performances of these antennas in the targeted body-centric scenarios (i.e., antennas mounted in the close proximity of the human body), have been analyzed with numerical simulations. The results, in terms of the signal fidelity, SAR, and radiation efficiency, show significant improvement in the case of the directional slot antenna design.

5.3 Channel Simulation and Measurement Methodology

5.3.1 Simulation of the Radio Propagation in Body-Centric Communication Scenarios

In Chapter 2, we have presented the propagation modeling methods and electrical characteristics of the human tissues, which are generally used for simulations of on-body or around-the-body propagation scenarios.

FDTD has several advantages that are useful for analyzing the UWB propagation environment around the human body. By transmitting a time domain Gaussian monocycle, for example, a wide range of frequencies is solved in only one simulation run, facilitating UWB propagation analysis.

On the other hand, compared to the low-frequency case discussed in Chapter 4, it is difficult to accurately recreate the full human body in a practical UWB simulation space, due to the complex shape and structure of the human body, with several different tissue layers. An anatomically correct model of a body is provided by the Visual Human project of the National Library of Medicine [44]. The mesh for this body model is accurate to within $5 \times 5 \times 5$ mm, allowing frequencies up to 6 GHz to be simulated correctly around the body.

Antenna models are generally not used in the full-body simulations. Accurate modeling of small UWB antennas would require a very fine grid resolution, making simulation over the entire body prohibitively complex. Rather, an electric field is generated directly by application of a voltage across one of the FDTD cells. This allows for investigation of signal path loss versus distance trends, but will not take into account losses due to the near fields of real antennas.

An example of a UWB propagation simulation setup is given in Figure 5.11. We will also use this example for our discussion on the body-area radio propagation phenomena in Section 5.4.3. All channel parameters are extracted from nine simulations, performed in planes separated by 4 cm along the z-axis of the torso (see left diagram). For each of these nine simulations, the transmitter is placed on the front of the torso. The electric field is observed in the x-y plane at several positions separated by 4 cm around the human torso (see right diagram). To increase the number of points for extracting channel statistics, observations that are one plane above and one plane below the transmitter are recorded in each simulation. In this way, a total of 570 datapoints are taken at various positions.

The radio channel parameters change depending on the position around the body. To easily describe this phenomenon, one can define three main angular regions, representing the front (0° to 60°), side (60° to 160°), and back (160° to 180°) locations on the body, as shown in Figure 5.11.

5.3.2 Measurement of the Radio Propagation in Body-Centric Communication Scenarios

Practical alternatives to the numerical simulations for the around-the-body channel investigations are the radio measurements. Unlike FDTD simulations that are limited by computational complexity, channel measurements can be used to capture

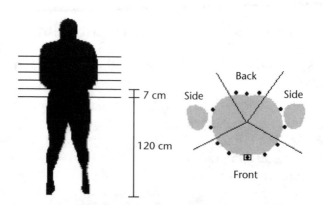

Figure 5.11 Body area simulation setup example, showing the positions of the transmitter (square on the front of the body) and the receiver (circles around the body). Measurements were taken in several planes, as shown on the left. Front, side, and back regions were analyzed separately. (*After:* [45, 46].)

and determine the influence of a larger radio environment on the around-the-body signal propagation. Furthermore, measurements inherently include the effects of the real UWB antennas, thus giving a more realistic set of results. Several different on-body antenna positions and scenarios can be applied in the measurements, illustrating possible body movements and potential on-body positions for the devices.

Typical radio channel measurements can be performed either with body phantom models [41, 47], or with real test persons [3, 41, 45, 46, 48, 49]. The more commonly used body phantoms are simplified replicas of the human body, and these simplifications can have a significant impact on the measured channel characteristics, especially at high frequencies and large fractional bandwidths, as in the case of UWB signals. Some of the body phantom types that can be used in radio propagation measurements have been described in Chapter 2.

The radio channel measurements with body phantoms are generally static investigations—that is, the body phantom (either its limbs or the entire phantom) is not moved in the given propagation environment during one channel sounding test. The radio channel investigations using a body phantom have the main advantages of good repeatability of the measurement results and easier availability for lengthy investigations.

With test persons, the results from the investigations can provide good statistical data for scenarios that are very close to real-life scenarios, in terms of device/terminal handling and user movements. In this case, it is more challenging to precisely reproduce the measurement results, due to motion of the body and difficulties controlling the exact position of the antenna and body.

As opposed to the investigations with a body phantom, the radio channel measurements with test users can also be dynamic investigations—that is, the user can move in the given propagation environment during channel sounding and the time-variant channel can be measured. In order to take advantage of this possibility, the measurement equipment has to accommodate the expected Doppler shifts in the radio channel estimated at the UWB upper frequency.

Traditionally for UWB radio channel measurements, due to the large fractional bandwidth required, a VNA is used, and the complex frequency transfer function of the channel (S_{21} parameter) is recorded for a large set of discrete frequencies. While this is appropriate for static UWB measurements, other solutions that allow higher sampling rates need to be employed for time-variant UWB channel investigations. One option is to sound the channel with a high bandwidth pulse, similar as in the FDTD simulations, and then detect and digitize the received signal [2, 50]. Another alternative is to use a time-domain channel sounder with an appropriate bandwidth and center frequency [47, 49, 51–53]. The time-domain radio sounding setups generally allow measurements on several parallel channels. Thus, more complex, ad hoc communication scenarios can be investigated, such as multiple-input multiple-output (MIMO), interference, relaying, and so forth. In addition to the aspects discussed above, there are several other channel measurement practicalities to be considered in the body-centric channel investigations, similar to the traditional, narrowband, and wideband channel investigations, such as transmit power, type of antennas, multiple (distributed) antenna elements setups, and so forth.

In terms of specific body-centric measurement equipment setup, the terminals and devices used as transmitters and receivers must allow the user relatively high freedom and natural movement. This implies the use of light, thin RF cables with good phase stability [48], or alternatively, the use of the RF-on-fiber setup for those experiments in which it is desirable to have more control over the radiation characteristics of the antennas [38, 41, 47]. Experiments show that the radio environment in the near range of the user (human body) can strongly influence the propagation channel [3, 38, 41, 46, 48, 54]. Thus, for isolating only the radio propagation channel characteristics due to the human body, the measurements have to be performed either in an anechoic room [7, 18], or in an environment with only distant radio scatterers [3, 46–48]. Although the indoor UWB radio propagation channel has been extensively studied and presented in the literature [55–57], the body-centric propagation phenomena are still not fully characterized in both time and time-delay domains [3, 41, 46–48, 56].

In the following sections, we give practical examples of body-centric UWB channel investigations with both static and dynamic users. We will highlight the most important aspects of the measurement methodologies used, and summarize the channel parameters possible to measure with a given experimental setup.

5.3.2.1 Static Channel Investigations

Figure 5.12 shows a typical VNA experimental setup to measure the static propagation channel near the body [3, 7, 41, 48, 58]. A VNA is used to measure the frequency transfer function between two antennas placed at various positions on a human body. The two antennas are connected to the VNA using low-loss coaxial cables. Due to the high path loss around the body, a wideband amplifier can be used to increase the signal-to-noise ratio in some of the antenna locations. To extract time-delay domain channel parameters, the VNA measurements are converted by first applying a frequency domain window (e.g., a Hamming window) to reduce sidelobes, and then applying an inverse Fourier transform. Before each set of measurements, the VNA is calibrated with the amplifier included in the chain, so that any deviation from the flat frequency response specification is removed.

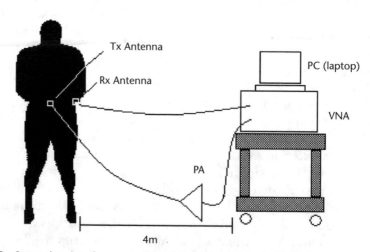

Figure 5.12 Setup showing the connection of the antennas to the VNA. A power amplifier can be used to increase the signal-to-noise ratio. A portable laptop computer can be used to control the VNA via a general purpose interface bus (GPIB).

The measurements were made either in an anechoic chamber [7, 58], in a large empty room [3, 41, 48] (so that reflections off the walls arrive later than the components diffracting around the body), or in a normal office room [58]. The around-the-body diffracted multipath components (MPCs) arrive at the receiver from 0.5 to 2 ns later, depending on the position on the body. The earliest reflected MPCs are due to the ground and arrived after 8 ns. Reflections off the rest of the surrounding radio environment arrive considerably later, and these MPCs can easily be identified and separated from propagation near the body.

Commercial low-profile UWB antennas [22] have been used for most of the body-centric measurements [3, 41, 45, 46, 48, 58]. These antennas were chosen since they represent the kind of form factor and profile requirements typical of comfortable body-worn sensor devices. Other investigations have been performed with larger antennas, such as the HSCA and the PICA [7, 18].

Examples of scenarios with antennas mounted on the body are shown in Figure 5.13. In Figure 5.13(a), the same setup is used as in the simulations presented in Figure 5.11. However, due to practical difficulties, fewer distances around the body can be analyzed in the measurements. For the investigations in the anechoic chamber, 22 different body scenarios and positions have been used in order to disclose the influence of the antenna characteristics on the propagation channel behavior [see Figure 5.13(b)]. Furthermore, the experiments showed that the distance between the body and the small-size antennas has a dramatic influence on the path loss, and should therefore be carefully controlled, as exemplified in Figure 5.14.

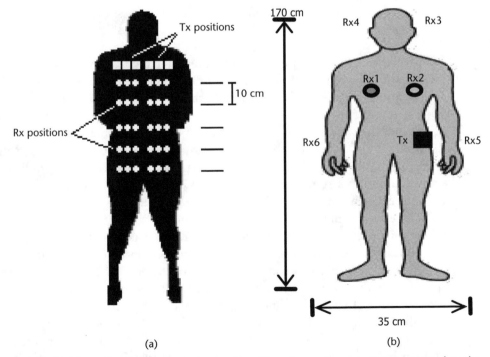

(a)   (b)

Figure 5.13 Example of measurement scenarios with various antenna positions mounted on the body. (a) Setup on the front of a body, used for simulations and measurements. (*From:* [46].) (b) Antenna positions for on-body channel characterization in anechoic chamber. (*After:* [7].) The boxes indicate transmitter positions, while the circles indicate receiver positions.

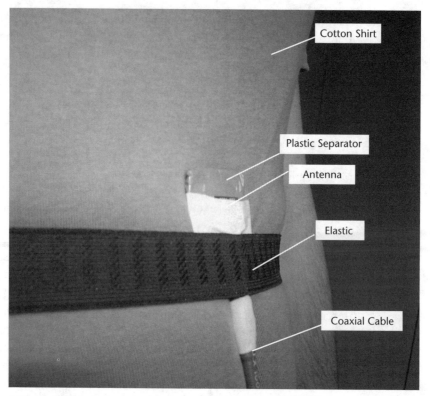

Figure 5.14 Example of a monopole antenna mounted on a body. A dielectric is used to control the separation from the body, and elastics are used to hold the antenna firmly in place. (*From:* [48].)

Measurement setups similar to those presented in Figure 5.13 have been used for investigating the effects of different arm movements and positions. These kind of investigations have been carried out in both frequency domain (see Figure 5.12) [1, 38, 41, 46], and in time domain (see Section 5.3.2.2) [47].

Rather than studying only the early MPCs due to diffraction around the body, the MPCs due to reflections off the radio environment arriving after the diffracting waves can be also investigated. In these scenarios, the structure of the radio environment is important. As an example, the indoor office layout used in the following discussions is given in Figure 5.15 [48].

The experimental procedure proposed in Figure 5.15 involves automatically taking measurements between antennas worn on a person at the various marked locations in the room. The transmitting antenna is placed on the front of the body, and the receiver is placed on either the front of the body (10 cm distance), the right side of the body (20 cm distance), or the back of the body (45 cm distance), as in Figure 5.11. In all cases, the antennas are separated from the body by 5 mm, as described in Figure 5.13. For these experiments, both the transmitting and the receiving antenna were worn on the body, at a height of 120 cm from the floor and 160 cm from the ceiling. At each of the nine locations in the room, 100-ns-long impulse responses are measured at 49 points arranged in a fixed-height 7×7 square grid with 5-cm spacing, covering 35×35cm [46, 56]. In general, the average power

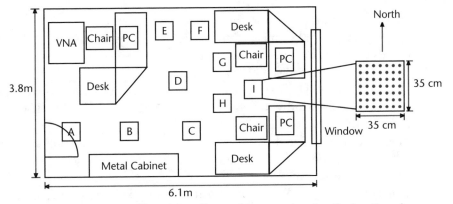

Figure 5.15 Example of floor plan of the office environment showing the location of measurement grids. Walls around the office are framed with metal studs and covered with wooden boards. The nine user locations used are labeled alphabetically from A to I [48].

obtained in each of the nine measurement grids is used to obtain large-scale signal statistics, while the variation over one measurement grid is used to obtain small-scale signal statistics.

Similar results can be obtained from dynamic channel measurements, where the test person actually moves in natural walking patterns in the given environment during the experiments. In this case, the small-scale signal statistics can be extracted from the measurements over a time span in which the large-scale structure of the channel can be considered constant.

A dynamic measurement procedure has the advantage of tracking changes in the channel structure due to the motion of nearby scatterers (such as the arms and legs), and provides estimates of the channel stationarity and variance in nonstationary radio propagation conditions.

5.3.2.2 Dynamic Channel Investigations

In this section, we present recent investigations of the body-centric radio channels using moving and walking users. These dynamic channel investigations focused on both personal area scenarios (i.e., distance ranges up to 10m), and body area scenarios (i.e., on-body distance ranges up to 1m). In order to obtain more general results, typical user scenarios and several indoor radio environments have been chosen. These results, similar to the traditional narrowband and wideband channel measurements, allow the analysis of the time-domain characteristics of the propagation phenomena.

In the case of moving terminals, the scattering environment "seen" by the antennas is also changing as a function of time and speed of the user. Given the large signal bandwidth, these changes can have a more significant influence on the UWB propagation channel, compared to the case of narrowband and wideband channels. Time-domain channel aspects can be directly estimated from the dynamic measurements using user movement tracks and normal walking patterns. In principle, the same time-domain and space-domain characteristics can be derived from static measurements, as described in Section 5.3.2.1. This assumes that the spatial sampling rate is high enough and the radio environment does not change significantly between the consecutive measurements, which are taken at significantly larger time intervals compared to the dynamic measurements.

The time-domain radio channel measurements presented here were conducted with a 4×4 MIMO UWB swept time delay correlation channel sounder [47, 49, 52]. The measurement bandwidth was 2.5 GHz, centered at 4.5 GHz. The radio channel sampling rate (time domain) was 20 Hz, which accommodates the expected channel change rate at the normal indoor walking speeds of 0.25 to 0.5 m/s (at 5 GHz). This setup is fast enough to extract small-scale channel characteristics. However, the time between subsequent measurements (50 ms) is too slow to track the evolution of the channel over a single UWB communication burst. The dedicated measurement setup allowed the full separation of all the 16 simultaneously measured radio links, and the measured channel impulse response data was compensated for the effects of all the system components except the antennas. The details for the measurement setup can be found in [47].[1]

The envisaged PAN scenarios include both mobile-to-mobile communications, between two moving users with handset-size devices, and mobile-to-stationary communications, between a moving user with a handset-size device and a stationary PC-laptop-size device [47]. In our following discussions, all these PAN scenarios are referred to as *PAN with mobile devices* or PAN-MD, and we only focus our analysis on the radio channel results for the mobile-to-mobile case.

In the PAN mobile-to-mobile scenarios, simultaneous measurements with four users were performed: two users with handsets as transmitters, and two users with handsets as receivers. The transmitter handsets, each with two antenna elements, were equipped with RF-on-fiber connections in order to improve the handling and mobility of the users. The distances between the transmitter and receiver handsets ranged from 1m to 6m. For reference, additional sets of static free-space measurements (i.e., without the movement along the routes and without user-proximity) have been performed, with the handsets mounted at the same height and orientation as in the user-proximity scenarios.

Figure 5.16 shows the handset use cases investigated in the body-centric mobile-to-mobile scenarios. Three main cases have been considered: handheld, PDA-held, and belt-mounted. The main difference between these scenarios is the user hand and body proximity in the near field of the antennas. In the handheld case, the antennas were mostly free, while in the PDA-held case, they were partly covered by the user's fingers and hand. In the belt-mounted case, the large dielectric body in the antenna near field significantly changes the monopole antenna radiation properties.

As already discussed in Section 5.3.2.1, several body-worn device positions in the case of the BAN scenarios are required in order to obtain statistically significant results. For the BAN channel investigations with users following a natural movement pattern, the body-worn devices have been used in five different scenarios: legs, arms, lower torso, upper torso, and two headset setups, as depicted in Figure 5.17 [47]. Each body-worn device had a directional slot antenna element as described in

---

1.   The documents are public, technical deliverables from the European IST-507102, "My Personal Adaptive Global NET (MAGNET)" research project, and can be downloaded from the official Web site of the project, http://www.ist-magnet.org.

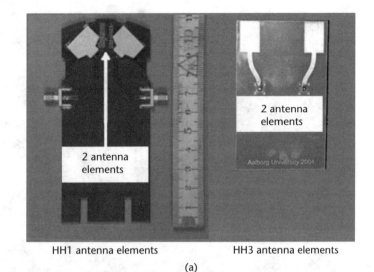

HH1 antenna elements   HH3 antenna elements

(a)

Hand-held   PDA-held   Belt-mounted

(b)

Figure 5.16   Handset scenarios for PAN-MD radio channel investigations: (a) dual antenna configuration in the handsets, and (b) handset use cases. (*From:* [47].)

Section 5.2, and was used as a transmitter, equipped with the RF-on-fiber setup in
order to maximize the accuracy of the obtained channel data [38]. For a receiver, a
handset-size device was used, equipped with two UWB monopole antennas similar
to the setup in the PAN-MD investigations [see Figure 5.16(a)].

For BAN investigations, reference measurements can be performed with a body
phantom. A cylindrical body phantom (without limbs) as described in [1, 41] was
used for reference measurements in representative locations (start, middle, and end)
of each user route [47].

The radio environment determines the main characteristics of the propagation
channel even for short-range communication systems. For the PAN-MD and BAN
scenarios presented here, the channel measurements have been performed in a laboratory environment, a corridor environment (confined area), and small- or
medium-size office environment. Figure 5.18 shows, as an example, the layout for
the laboratory or professional environment, and the corresponding user movement
paths. The test users in the PAN-MD and BAN scenarios were moving at normal
walking speeds, between 0.25 and 0.5 m/s. Although the walking paths were prede-
termined, the test users used their own natural walking patterns. The walking speed
has been determined by the average walking time on a given path.

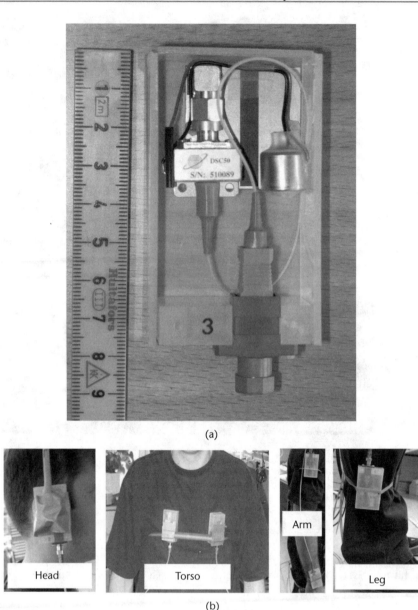

(a)

Head   Torso   Arm   Leg

(b)

Figure 5.17 Body-worn scenarios for BAN radio channel investigations with dynamic users: (a) body-worn device and antenna, and (b) body-worn device use cases. (*From:* [47].)

5.3.3 Concluding Remarks

In this section, we have presented the simulations and measurement methodologies currently used for the UWB body-centric radio propagation investigations. Examples of simulation methods and practical measurement setups have been reviewed, highlighting the importance of a careful design of the "on-body" experiments: on-body classifications, antenna types and their position relative to the human body (or their model), large-scale radio environment analysis, user dynamics, and so forth. While simulation can be generally performed with numerical tools, such as FDTD, and allow a very precise body structure modeling, important antenna radia-

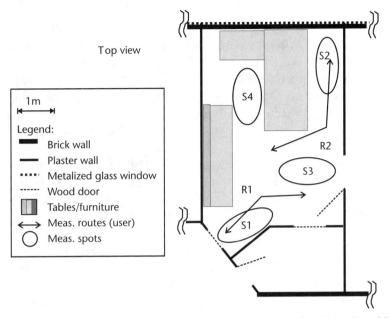

Figure 5.18   Example layout for the office-type radio environment in the PAN-MD and BAN channel investigations with walking users. (*From:* [47].)

tion characteristics are difficult to include. On the other hand, appropriately designed measurements can provide accurate parameters of the complex on-body propagation channel, including realistic radio environments, various antenna designs, and user dynamics.

The different signal analysis procedures have been also presented, and form the basis of the channel characterization and modeling results presented in Section 5.4.

5.4   Channel Characterization and Modeling

The results from the UWB radio channel investigations, simulations, and measurements, described in Section 5.3 are used throughout this section in order to describe and explain the radio propagation mechanism in body-centric PAN and BAN scenarios. From the radio propagation point of view, the main characteristic of these body-centric scenarios is that both the transmitter and receiver side of the communication link are in a relatively dense, scattering environment, including the human body proximity. This also represents the main novelty in the channel investigations and models presented herein, relative to the majority of existing UWB PAN channel models proposed in the literature [53, 55–57].

Although the parameters for a modified Saleh-Valenzuela (SV) UWB model [55, 56, 59] have been extracted and analyzed for the results presented in the following, this parameterization can still be an ambiguous procedure. For example, definitions of the channel feature signal clusters, and the associated thresholds (above the noise level or measured from the impulse response peak) used to extract them, can differ greatly between researchers.

5.4.1 General Aspects

The general description of the parameters for the modified SV UWB channel model can be found in [55, 56]. Here, we use the same notations as in the IEEE 802.15.3a, and highlight only the differences introduced. In this section, we present a quantitative description of the average channel power delay profile and of the associated large-scale parameters. A more detailed analysis can be found in [3, 7, 18, 45, 47, 48, 52, 54]. For convenience, the parameters and notations used in this section are listed in Table 5.2.

A convenient characterization of the UWB multipath propagation channels is the discrete-time impulse response model [50, 53, 57, 60]. In this model, the time axis is divided into small time intervals called "bins." The bin size is generally chosen to be the resolution of the measurements and simulations, since two paths arriving within a bin cannot be resolved. In our case, we assume that each bin contains a single MPC with amplitude corresponding to the energy in that bin. In the investigations where the CLEAN algorithm has been used to extract the channel impulse response parameters, the final model is a discrete-time impulse response with a bin size corresponding to the original measurement bandwidth. This bin size was obtained by applying a CLEAN algorithm with a "dirty beam" sampled at the original time-delay resolution of the channel.

There were three main characteristics identified that differentiate the analyzed body-centric channels from the traditional UWB LOS/NLOS channels reported in the literature [55, 56]. The first main characteristic is the dual slope power decay in the time-delay domain for the average cluster peak power with two different and environment-dependent decay factors, Γ_1 and Γ_2, below and above an excess delay threshold, T_{env}, respectively:

$$\propto \begin{cases} \exp(-T/\Gamma_1) & T < T_{env} \\ \exp(-T/\Gamma_2) & T \geq T_{env} \end{cases} \tag{5.4a}$$

Table 5.2 Main Channel Parameters and Notations Used for the UWB PAN-MD and BAN Channel Characterization and Modeling

Parameter (Units)	Description
(A) PDP	(Average) power delay profile
T (ns)	Signal cluster arrival time
T_{env} (ns)	Time-delay threshold within each PDP
Λ (1/ns)	Signal cluster arrival rate
Γ_1, Γ_2 (dB/ns)	Average intercluster peak power decay factor(s)
γ (dB/ns)	Average intracluster power decay factor
σ_{c1}, σ_{c2}	STD of cluster(s) log-normal fading
σ_s	STD of wideband power log-normal large-scale fading
τ_{rms}	RMS delay spread
$W90$	90% energy time-delay window
N_p	Number of significant multipath components with −10 dB of the largest component in each PDP

and

$$\Gamma_1 < \Gamma_2 \tag{5.4b}$$

A similar modeling approach was proposed for the IEEE 802.15.4a WPAN channel, with a linear increase of the intracluster power decay rates with excess delay, in dense/industrial radio scattering environments [56]. In our channel model proposals, the intercluster power decay follows (5.4), while the intracluster power decay is modeled as exponential with a constant decay rate, γ, within the entire impulse responses.

From independent experimental investigations (static and dynamic), the excess delay threshold, T_{env}, was found to vary in the range from 25 ns, as reported for the PAN-MD and BAN scenarios in [47, 52], up to 40 ns, reported for the BAN scenarios investigated in [46, 48]. In both investigations, the $\Gamma_1 < \Gamma_2$ relationship has been disclosed, although the difference between the two decay factors was different in the PAN-MD and BAN scenarios. Furthermore, this dual-slope phenomenon was observed also in the free space (i.e., without user proximity) PAN-MD scenarios.

A possible explanation for the observed dual-slope time-delay decay of the PDP expressed in (5.4) resides in the inhomogeneous scattering around the antennas, due to the interaction between the antenna radiation patterns, which are far different from the omnidirectional characteristics used in earlier UWB investigations, and the dense scattering environment in which both transmit and receive antennas are immersed. This dual-slope characteristic of the PDP is still a topic for further investigation.

In the PAN-MD scenarios [47, 52], due to the dominant horizontal orientation of the antenna elements (monopoles, see Figure 5.16), the shortest signal paths transmitted or received at low angles relative to the direct line between the antennas, are significantly attenuated by the antenna radiation patterns, even in the free-space scenarios. The next two to three signal clusters up to 25-ns excess delay (7.5-m excess distance) are transmitted and received at larger angles, and although they are more attenuated due to scattering, they are also radiated and received with less antenna attenuation compared to the first signal clusters. This combination yields the relatively low-signal attenuation rate determined between the first three to four signal clusters in the PAN-MD scenarios. At higher excess delays, beyond 25 ns, the influence of the antenna radiation patterns is lower, due to the propagation through multiple scattering, and a high cluster attenuation rate is observed. In the user-proximity scenarios, although the antenna radiation characteristics are changed slightly, the general signal cluster attenuation trend is the same.

In the BAN scenarios, the antenna radiation patterns exhibit certain directivity characteristics, either due to the proximity of the body, as in the case of the monopole antennas (see Figure 5.14) [45, 46], or because of their special design, as in the case of the directive slot antennas (see Figure 5.5) [19, 47]. The dual-slope time-delay decay was predominantly determined when both transmit and receive antennas were placed on the same side (front) of the body. The basic explanation given for the PAN-MD scenarios is plausible also for the BAN scenarios, with the main difference that the first signal cluster is due to diffraction around the body, and the next two to three signal clusters are generally received due to scattering from the ceiling and floor areas in the near range of the user.

The second main characteristic is the difference in cluster fading factors for the two time-delay regions in the user-proximity scenarios. In all our dynamic scenarios, PAN-MD and BAN, the signal clusters were found to follow a log-normal power distribution in time-domain, similar to IEEE 802.15.3a/4a models [55, 56]. However, the standard deviation values for the cluster fading were found to be generally higher than proposed in the IEEE 802.15.3a/4a models. In the mobile-to- mobile PAN-MD scenarios, higher cluster fading was detected in the first time-delay region, $T < T_{env}$, compared to the second region, $T \geq T_{env}$. This radio channel characteristic is discussed further in Section 5.4.2.

A third important aspect that needs to be analyzed for the radio channels with moving users is the time/space-domain channel decorrelation time/distance. In the IEEE 802.15.3a/4a, PAN and BAN channel models time-domain aspects are not included [55, 56]. In the UWB PAN-MD and BAN channel investigations with moving and walking users, the channel stationarity was analyzed with respect to the large-scale wideband power decorrelation time/distance. Within this decorrelation time interval, the channel was also found to exhibit average similar time-delay signal clustering characteristics (power and time delay) [47, 49].

5.4.2 Personal Area Network Scenarios

The UWB mobile-to-mobile communication scenarios have been recently investigated, and a channel model proposal is available [47, 49] as an enhancement to the existing channel models [53, 55–57, 61].

In this section, we briefly summarize the main results from these UWB mobile-to-mobile channel investigations, described in Section 5.3.2.2. The channel impulse responses corresponding to all user movement routes in a given environment have been combined and analyzed together in order to obtain statistics over all user locations along the routes. Furthermore, to obtain statistics over all possible user device scenarios, handheld, PDA-held, and belt-mounted (see Figure 5.16) in a given environment, these datasets have been combined and analyzed together. The latter averaging was possible because the results showed relatively small variations of the extracted channel parameters over the different user device scenarios. A customized subtractive clustering algorithm, together with a simple sensor-CLEAN processing for signal ray extraction, have been used for the extraction of the time-delay parameters [47]. The signal-to-noise ratio (SNR) range in the measurement data was from 15 to 25 dB.

For these PAN-MD scenarios, we analyze and present only the following main characteristics, which can be used to enhance the existing static, time-invariant UWB PAN channel models:

- Large-scale time and space channel decorrelation aspects;
- Fading statistics of the individual signal clusters and of the total wideband power;
- Signal cluster power decay and signal cluster arrival rate.

The detailed analysis of the other channel parameters (e.g., signal ray statistics) can be found in [47, 49].

5.4.2.1   Large- and Small-Scale Fading

In order to identify the small-scale and large-scale fading processes in the PAN-MD channel, the channel stationarity has been analyzed first on representative datasets. The autocorrelation function of the wideband power has been used to test the channel stationarity over different time spans.

Figure 5.19 shows the average autocorrelation functions of the wideband power, determined for the user-proximity scenarios in the three environments investigated.[2] As mentioned earlier, these environments differ in the spatial density and distribution of the radio scatterers, which can also be seen on the autocorrelation curves in Figure 5.19.

With the channel sampling rate of 20 Hz and the user speed of 0.5 m/s, consecutive channel samples (spaced 2.5 cm apart) can be considered that are fully decorrelated in terms of wideband power. Thus, the autocorrelation curves in Figure 5.19 are dominated by the large-scale signal fading. Based on these results, it is concluded that a large-scale fading decorrelation time interval of 0.5 second can be considered an average for all PAN-MD scenarios (at 0.7 correlation level). This time interval corresponds to 10 consecutive measured channel impulse responses and to an average spatial decorrelation distance (along the user movement track) of approximately 0.25m. Within this time (or distance) interval, the small-scale signal fading can be considered as a stationary process. The 10 consecutive channel

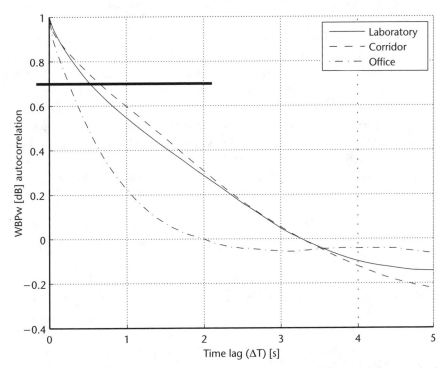

Figure 5.19   Wideband power autocorrelation functions for the mobile-to-mobile scenarios in the investigated environments.

2.   The logarithm of the normalized signal power has been used in order to render the combination of the small-scale and large-scale signal variations an additive process.

impulse responses also have similar time-delay clustering characteristics, which is a direct result of a similar distribution of the main radio scattering sources (as "seen" by the antennas) during the 0.5-second time interval.

The APDPs are calculated using 10 consecutive measured channel impulse responses. Although the correct Doppler sampling rate has been used, this low number of impulse responses did not allow the extraction of reliable small-scale statistics within each APDP. For the modeling of the small-scale channel characteristics, within the 0.5-second time and 0.25-m distance interval, the distributions types in the IEEE802.15.3/4a channel model proposal could be used, possibly with a different set of parameters [55, 56].

Table 5.3 lists the parameters for large-scale signal statistics of the signal clusters and the wideband power in the investigated environments. In the user-proximity scenarios (US), the average wideband power trend along a given user route is removed when calculating the standard deviations values listed in Table 5.3. In the static free-space measurements (FS), only a low number of spatial positions along the routes have been used, so that the large-scale wideband power fading statistics could not be reliably extracted. The signal cluster fading in FS scenarios could be analyzed more reliably when using the total of four to seven clusters detected per APDP.

In the time and space domain, the cluster's peak power distribution was determined to be close to log-normal, and independent fading was observed for the extracted clusters, similar to earlier proposals. The cluster fading was extracted relative to the averaged cluster peak power levels over all average PDPs along the user route or free-space locations.

In the user-proximity scenarios with two moving users, for the clusters in the first time-delay region, the fading standard deviation is significantly higher than in the second time-delay region, $\sigma_{c1} > \sigma_{c2}$ in Table 5.3. This is the effect of the simultaneous dual dynamics in the radio channel, with two users moving, and both inducing independent log-normal large-scale fading. The signal cluster fading in the first delay region, mostly determined by the body proximity, is stronger than that determined in the IEEE 802.15.4a NLOS PAN channel [56]. In general, the cluster fading standard deviation values in the PAN-MD free-space scenarios are in the same range as specified for the IEEE 802.15.4a LOS PAN channel [56].

5.4.2.2   Power Delay Profile

Figure 5.20 shows examples of APDPs and the detected signal clusters corresponding to the determined average large-scale decorrelation time interval of 0.5 second (10 consecutive impulse responses) in the mobile-to-mobile scenarios. Table 5.4 lists the main time-delay channel parameters extracted from the APDPs.

Table 5.3 Large-Scale Fading Parameters for the Mobile-to-Mobile Radio Channels in the Free-Space (FS) and User-Proximity (US) Scenarios [47, 49]

		Laboratory	Corridor	Office
Cluster fading σ_{c1} & σ_{c2} (dB)	FS	4.6	4.1	3.1
	US	6.9 & 4.8	6.3 & 3.3	4.7 & 2.8
Wideband power fading σ_s (dB)		3.6	4.4	2.7

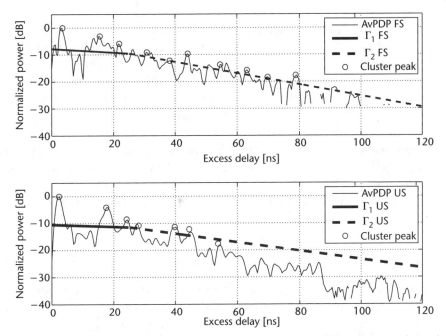

Figure 5.20 Example of measured average channel power delay profiles and detected signal cluster peaks in the laboratory environment for free-space (FS) and user-proximity (US) mobile-to-mobile scenarios.

Table 5.4 Main Time-Delay Parameters for the Mobile-to-Mobile Radio Channels in the Free-Space (FS) and User-Proximity (US) Scenarios [47, 49]

		Laboratory	*Corridor*	*Office*
Intercluster power decay	FS	0.07 & 0.21	0.07 & 0.20	0.08 & 0.27
Γ_1 & Γ_2 (dB/ns)	US	0.04 & 0.16	0.15 & 0.18	0.21 & 0.26
Intracluster power decay	FS	1.65	1.64	1.59
γ (dB/ns)	US	1.65	1.46	1.82
Cluster arrival rate	FS	0.230	0.216	0.229
Λ (1/ns)	US	0.215	0.170	0.275

A possible explanation for the lower cluster power decay factor in the first excess delay region up to $T_{env} = 25$ ns [see (5.4)] has been already given in Section 5.4.1. It is worth noting the differences in terms of power decay factors between the three investigated radio environments. In free-space conditions, there is little influence of the radio environment on this channel parameter. However, in the user-proximity scenarios, the power decay factor increases when the radio environment becomes increasingly confined (laboratory, corridor, and office, in that order), indicating a significant interaction between the user body, antennas, and environment.

A comparison with the IEEE 802.15.3a [55] channel model proposals highlights the main difference in the signal clustering in the proposed PAN-MD model.

On average, 3.3 times higher signal cluster arrival rates and 3.5 times higher signal ray decay factors within the clusters have been determined in the PAN-MD scenarios, compared to the CM3/CM4 NLOS PAN models. These differences are mainly due to the radiation patterns and orientations of the UWB antennas used. The antennas used in the IEEE investigations had vertical orientation and an omnidirectional pattern over the entire frequency band, while this was not the case for the antennas in the PAN-MD measurements. An important factor in the extraction of these parameters is also the definition of the feature signal cluster. In the processing of the PAN-MD data, the clusters were extracted based on a power-delay clustering algorithm, while they were visually identified for the IEEE channel model proposal. These two methods can inherently yield the identification of different signal clusters sets within the same average power delay profile.

5.4.2.3 Time/Space Variant Channel Modeling Aspects

The detailed PAN-MD channel model parameters and description are available in [47, 49]. The proposed channel model for the scenarios with dynamic users uses the main parameters listed in Tables 5.3 and 5.4. A time-domain impulse response generation procedure is proposed, based on the determined large-scale correlation properties presented in Section 5.4.2.1. Consecutive channel realizations in one decorrelation time interval of 0.5 second have the same large-scale fading factor and the same average time-delay clustering characteristics. Due to the movements of the antennas and users, when the signal shadowing is changed, the signal time-delay clustering is also changed. The correlation level between consecutive large-scale wideband power fading factors is extracted from the curves given in Figure 5.19, or alternatively, can be considered completely independent.

A more precise time and space-variant modeling of the signal clustering has been proposed in [62], based on a hybrid statistical and quasideterministic environment modeling. The topic is still the subject of further investigations, especially for the PAN user-proximity scenarios.

In Table 5.5, we give a comparison between the statistics of the most widely used channel parameters, obtained from PAN-MD measurements and simulations. All of the delay parameters were determined to exhibit a normal distribution, with a mean ($<>$) and standard deviation (STD) determined over all the user routes, terminal types, and scenarios in a given environment. The comparison in Table 5.5 shows a good match between measurements and simulations for the wideband power

Table 5.5 Comparison of Channel Parameters Extracted from Measurements and Simulations in User-Proximity Mobile-to-Mobile Scenarios [47, 49]

		Laboratory	Corridor	Office
Wideband power fading σ_s (dB)	Measured	3.6	4.4	2.7
	Simulated	3.7	4.3	2.3
RMS delay spread τ_{rms} (ns)$<>$/STD	Measured	24/7	25/8	15/6
	Simulated	25/10	25/11	15/5
90% energy window $W90$ (ns)$<>$/STD	Measured	52/15	57/17	30/10
	Simulated	62/32	59/30	37/15

shadowing and the average RMS delay spread. While the RMS delay spread values are mainly determined by the signal clustering, the 90% energy window parameter is much more sensitive to the detailed signal power distribution in the channel impulse response, a structure which is generally difficult to match in simulations. The relatively large STD values in the simulations are due to the variable length impulse response generation procedure.

5.4.3   Body Area Network Scenarios

The UWB radio propagation around the human body is a complex phenomenon, although it takes place over only very short distance ranges. For communication between two devices placed on the human body, transmitted signals can arrive at the receiver in three ways:

- Propagation through the body;
- Diffraction around the body;
- Reflections from nearby scatterers in the radio environment.

This section is structured as follows. Section 5.4.3.1 focuses only on the propagation around the body, while Section 5.4.3.2 extends these results to include the influence of the user's arm movements in terms of wideband power fluctuations. The influence of several indoor environments is presented in Section 5.4.3.3, based on results available from both static (frequency-domain) and dynamic (time-domain) channel investigations. Finally, based on the results, a complete channel model is proposed in Section 5.4.3.4.

5.4.3.1   Propagation Around Body

Signals in the gigahertz frequency range diffracting around the body attenuate due to absorption by human tissue. In addition, the original transmitted signal spreads out in time, due to the frequency-dependent dispersion by the antenna-body system. This attenuation and signal spreading likely depends on a number of random factors, including the curvature of the body, the exact position of the antennas, the position of the arms, the type of materials along the various signal paths, and so forth.

In the following sections, we describe the path loss, the power delay profile, and time-delay parameters, using several investigations reported in the literature [3, 7, 18, 41, 45, 46, 48, 54]. Since signal paths near the body are independent of the surrounding environment, the results presented in this section are generic, and can be applied to any typical indoor or outdoor scenario.

Path Loss   Simulation results indicate that very little signal propagation takes place through the body in the gigahertz frequency range. Instead, the radio waves are diffracting around the torso. Therefore, in calculating the path loss, we propose to measure the distance around the perimeter of the body [45]. Based on the results from FDTD simulations described in Section 5.3.1, Figure 5.21 shows an example of an electric field snapshot around the torso in the horizontal plane, taken 5 ns after transmitting a pulse.

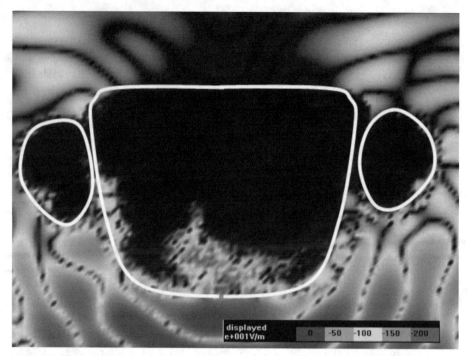

displayed
e+001V/m   0   -50 -100 -150 -200

Figure 5.21   Simulated electromagnetic field snapshot around the torso. Different colors represent the magnitude of electric field. (*From:* [45].)

Figure 5.22(a) shows the path loss versus distance trend extracted from the numerical simulations [45], while Figure 5.22(b) shows the same plot extracted from the measurement setup described in Section 5.3.2 [46]. To allow for easy comparison with measured results, the simulated path loss is shifted so that it equals the mean measured path loss at a reference distance of 0.1m, and the distances are measured around the perimeter of the body. Clearly, the path loss increases with distance, as expected, and there is a large variance around the mean path loss. These simulations results do not incorporate losses due to the antenna, since applying a voltage across an FDTD cell generates the electric field.

The measured on-body path loss strongly depends on the radiation patterns of the antennas used. To exemplify this result, Figure 5.23 presents the path loss values measured in the anechoic room and the modeled path loss curves as a function of distance, for the investigations using a HSCA and a PICA [7, 18].

The far-field path loss is usually best modeled with the empirical power decay law:

$$P[\text{dB}] = P[\text{dB}] + 10n \log_{10}\left(\frac{d}{d_0}\right) \tag{5.5}$$

where n is the path loss exponent, d is the distance from the antenna, and P_0 is the path loss at the reference distance d_0. The parameters of this path loss model extracted from the simulation and measurements are listed in Table 5.6.

The nonreflecting environment in the anechoic chamber leads to the high exponent of path losses [7, 18]. However, for PICA, good omnidirectional radiation

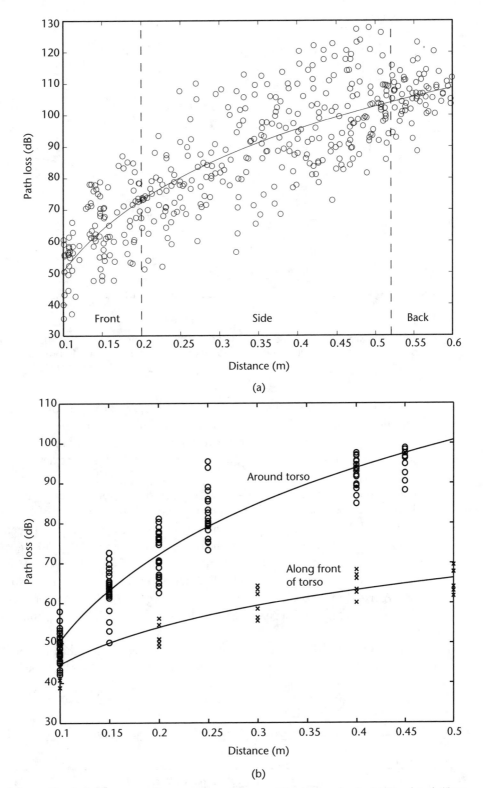

Figure 5.22   Path loss versus distance measured around the human body. (a) Simulated. (*From:* [45].) (b) Measured. Circles and crosses indicate individual measurements. The path loss model is obtained by a best-fit procedure. (*After:* [45, 46, 48].)

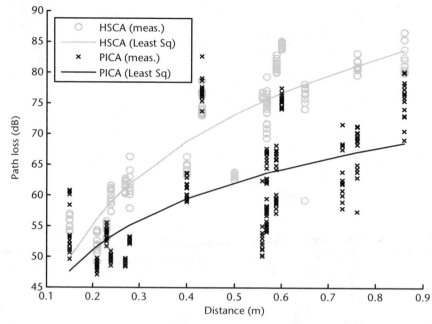

Figure 5.23 Measured and modeled path loss for the HSCA and the PICA versus distance measured around the human body. The path loss model is obtained by a best-fit procedure.

across the UWB range results in the lower loss exponent value, due to diffraction around the human body. In addition to the contribution to channel path loss by distance changes and different body positions, another important factor is the orientation of the antennas. For example, an additional 20-dB loss was determined when two PICAs were orthogonally oriented on the body [7, 18].

Compared with free space ($n = 2$), the path loss exponent near the body in the gigahertz range is much higher. For transmission along the front of the body, the index $n = 3.3$. For NLOS transmission around the body, n ranges between 7.2 and 7.4, which is much higher than in typical wireless systems. These higher path loss exponents are expected, since the propagation mechanism relies only on diffraction in the shadowed regions of the body. Furthermore, the path loss results for HSCA and PICA listed in Table 5.6 do not separate between LOS (along the body) and NLOS (around the body) scenarios, so their estimated path loss exponent is significantly different.

The mean path loss value depends also on the separation distance between the antenna and the body, due mainly to antenna mismatches, as described in Section

Table 5.6 Path Loss Model Parameters Obtained from Simulations and Measurements [7, 18, 45, 46]

Parameter in (5.5)	Simulated [45]	Measured [46]		Measured [7, 18]	
	Around Body	Around Body	Front of Body	HSCA	PICA
n	7.5	7.4	3.3	3.9	2.6
d_0 (m)	0.1	0.1	0.1	1.0	1.0
P_0 (dB)	—	50.5	44.5	86.5	70.3

5.2. The reference path loss at 0.1m given in Table 5.6 is for a body separation distance of 5 mm [45, 46]. For monopole antennas, this reference parameter was found to vary between ±6 dB, depending on the separation distance, generally increasing when the antenna is placed closer to the body.

To determine the signal amplitude distribution, the large-scale path loss (the line from Figure 5.22) was removed. The log-normal and Suzuki models provided a reasonable fit, while other candidate distributions could be rejected, based on visual inspection and goodness-of-fit tests. Therefore, the use of a log-normal distribution is recommended, since it is much easier to implement in practical channel simulators.

Empirical and theoretical log-normal fit distributions for consecutive time-delay bins are reproduced in Figure 5.24. Marginally higher variances were observed on the side of the body (~ 7 dB) compared to the front of the body (~ 6 dB), and may be due to reflections off the arms and shoulders. Smaller variances were also observed on the back of the body (~ 5 dB), possibly due to both components diffracting around the torso.

The log-normal distributions describing the energy in each delay bin were found to be correlated. Log-amplitude correlations in the range from 0.5 to 0.9 were measured in the first three (up to 1.5 ns) time-delay bins, regardless of the receiver position. The correlation coefficient decreases with increasing delay between bins. Physical phenomena that may contribute to correlated bins include the frequency dependency of human tissues, and the antenna and body system. Furthermore, multipath components due to reflections off the body and arms may be correlated,

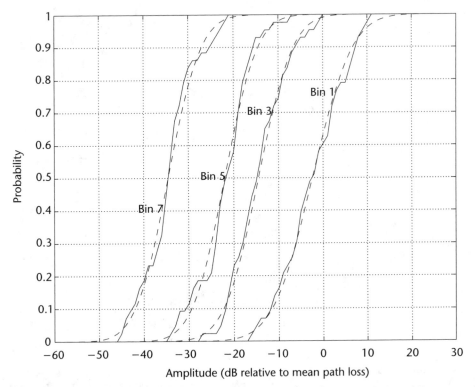

Figure 5.24   Example of amplitude cumulative density functions for Bins 1, 3, 5, and 7, measured along the side of the body, using 0.5-ns bins.

due to short path lengths, a natural symmetry of the body, and overlapping path trajectories, especially in the vicinity of the transmitter and receiver.

From a comparison of the results obtained in the simulations and measurements, we conclude that FDTD simulations are useful for extracting average path loss and power delay profile trends, but may not be suited for estimating precise statistical information describing how these parameters vary without very sophisticated numerical techniques or body and antenna models.

Impulse Response and Average Power Delay Profile   Figure 5.25 presents impulse responses for two on-body scenarios in the anechoic chamber investigations with the HSCA and PICA [7, 18]. It can be seen that most energy is received via the direct path, with some multipath reflections at the later time. The changes in channel characteristics due to different antenna radiation characteristics can be also found in these figures. The main dissimilarity between antennas is that more strong echoes and ringing effects appear in the PICA case. This can be explained by the fact that the PICA has a narrower bandwidth and a greater number of resonance frequencies within the measured band, which increases the signal spread.

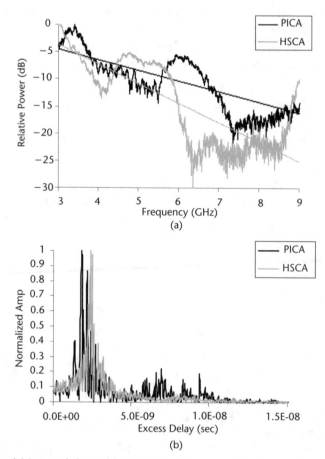

Figure 5.25   (a, c) Measured channel frequency responses; and (b, d) impulse responses with HSCA and PICA for two on-body scenarios: (a, b) scenario with receiver on the right side of the head with body standing still, and (c, d) scenario with receiver on the left wrist with arm stretched aside. (*From:* [18]. © 2006 IEE. Reprinted with permission.)

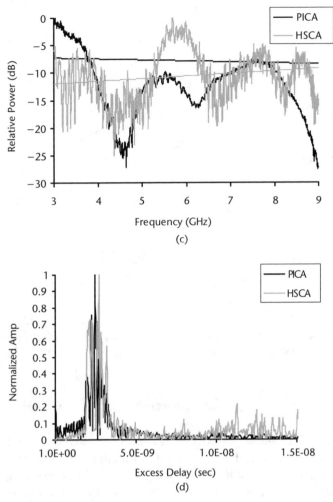

Figure 5.25   (continued.)

Figure 5.26 only presents an example of measured APDP [3, 48]. Similar results are obtained using FDTD simulations [45, 46]. The energy of subsequent time-delay bins can be well modeled with an exponential law as a function of excess delay. However, the FDTD simulations did not incorporate real antennas, and the observed longer impulse responses in the measurements can be explained with near-field effects and/or the convolution of the antenna response with the propagation channel response.

In [7, 18], the RMS delay spread and mean excess delay parameters have been calculated for an impulse response length of 80 ns measured in the anechoic room. Figure 5.13(b) shows the antenna scenarios investigated. RMS delay spread values ranging from 1 to 7 ns were determined for various antenna types and positions around the body. However, the investigations with small, low-profile monopole antennas yielded RMS delay spread values only in the range from 0.1 to 0.8 ns for various on-body device positions, as reported in [41, 48, 54]. The approximately one order of magnitude difference is clearly due to different antenna radiation char-

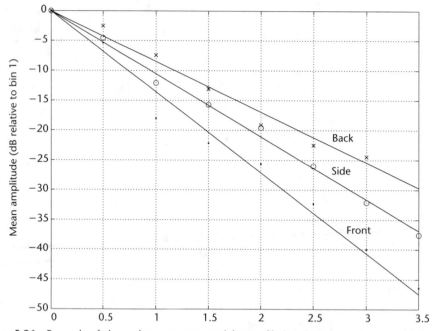

Figure 5.26 Example of channel average power delay profile in a BAN scenario. Measured decreases in mean amplitude over successive 0.5-ns bins relative to the first bin.

acteristics, underlying the importance of antenna design for specific BAN applications.

Figure 5.27 shows the RMS delay spread versus on-body distance for both measurement cases, using HSCA and PICA antennas. This distance relationship is also consistent with the results reporting longer impulse response on the back and side of the body compared with the front of the body [48]. The cumulative distributions of the RMS delay spread and mean excess delay parameters showed a good fit to normal distributions around the average RMS delay spread [7, 18].

As expected, the mean delay due to propagation link between the transmitter and the receiver is the highest where the NLOS channel and the propagation around human body on the surface (i.e., creeping waves) are the main propagation channels. For the cases where both antennas are placed on the same side of the human body, but still in NLOS conditions, the mean RMS delay spread measured with the HSCA was smaller than the RMS delay spread measured with the PICA. This can be related to the placement of both antennas, since stronger surface waves are launched with the HSCA. For other links, where LOS components (e.g., free-space waves) are dominant, the PICA produced a lower delay spread.

5.4.3.2   The Influence of the Motion of the Arms

Channel measurements indicate that the position and/or movement of the arms can significantly change the mean signal path loss between the transmitter and receiver [3, 38, 41, 46, 54]. For example, when the transmitter is placed on the belt and the receiver is placed near shoulder height, a drop in received power in a range from 5 to 20 dB is recorded when the arms are folded across the chest between the two anten-

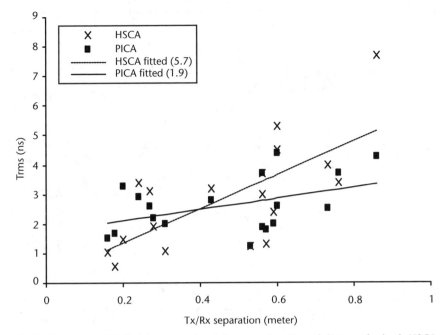

Figure 5.27 Example of RMS delay spread variation as a function of distance for both HSCA and PICA in a BAN scenario (*From:* [7].)

nas. The investigations with transmit and receive antennas placed either on the torso or arms showed that wideband power fluctuations due to the arm positions are in the range of ±15 dB from the median values.

Investigations also show that arm motions can significantly influence the fluctuations of the received signal levels. Figure 5.28 shows an example of measured received energy fluctuation due to arm movements, compared with the fluctuation determined with a motionless body [3, 46]. As expected, a walking motion of the arms produces measurable fluctuations if the receiver is placed on the side or back of the body. When both the transmitter and receiver are on the front of the body, the arms are too far away and do not significantly influence the received power level.

Analysis of the data indicates that the variation of the received energy around the mean path loss due to typical arm motions is better modeled by distributions such as the log-normal and Nakagami-m distributions, rather than with a simple Rayleigh model [3, 46]. This is likely due to a combination of two effects. First, the arms by themselves do not produce enough random phase signal paths to justify the Rayleigh distribution. Second, the resolution of a UWB system is very high, so there are not many irresolvable multipath components. Based on these results, we can conclude that a UWB BAN system design needs to account for the possibility of deep fades due to arm and body movements, especially if the receiver is worn so that body motions can shadow the receiver from the transmitter.

5.4.3.3 The Influence of the Indoor Environment

The signal propagation around the human body discussed in Sections 5.4.3.1 and 5.4.3.2 is only one component of the total BAN propagation channel. The

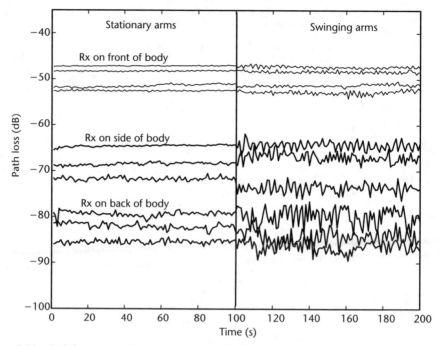

Figure 5.28   Path loss versus time in measurements taken at several receiver positions (front, side, back) while the user is motionless (first 100 seconds), compared with when the user is walking in place (final 100 seconds).

multipath radio environment in the near range of the users has also a significant impact on the BAN propagation channel, and needs to be included in the channel characterization and modeling [38, 41, 47, 48, 54].

In this section, the results from the static measurements in frequency domain, as discussed in Section 5.3.2.1, are used to highlight the differences between the extracted channel parameters for different antenna positions on the body [3, 46]. From the dynamic channel measurements in time domain, as discussed in Section 5.3.2.2, the same channel parameters are averaged over similar antenna positions on the body, and these average values are compared within different indoor environments [47].

These static and dynamic channel investigations have been performed independently, in buildings with different structures, with different antenna types, and using different data processing procedures. Thus, the results presented in the following sections are a very good indication of the variability of the extracted channel parameters, not only due to indoor radio propagation environments and on-body positions, but also due to the experimental procedures used.

Large-Scale and Small-Scale Fading   The variation of the total signal energy received depends both on the position of the antennas on the body and the location of the body in the room. This large-scale fading is commonly modeled with a log-normal distribution. Table 5.7 shows large-scale wideband power distribution measured in the static and dynamic BAN channel investigations. Different relative positions on the body are considered in the two measurements, due to the different locations of the main unit: the front of the torso for the static measurements, and the lower right side of the torso for the dynamic measurements.

In terms of wideband power levels, significant deviations can be observed between the results in the four considered environments. We concluded that these phenomena are due mainly to the *directional antennas* used in the body-worn devices in the case of the dynamic measurements [47]. The determined standard deviations of the large-scale log-normal fading agree well between the static and dynamic measurements, averaging less than 2 dB.

No significant variation of the signal cluster fading factors in the dynamic measurements was determined between various body-worn device positions. Thus in Table 5.7, only the average values are given for each radio environment. The average standard deviation of the log-normal cluster fading is slightly higher in the dynamic measurements, compared to the static measurements. This result can also be explained with a greater influence in the radio scattering environment due to the directional antennas used [47].

In addition, small-scale fading in each delay bin is observed, due to small changes in the user position. A statistical analysis from [46, 48] indicates that a log-normal or Nakagami-m distribution provides a plausible model to describe this phenomenon, and is consistent with other UWB indoor propagation studies [50, 55, 56].

Power Delay Profile   Figure 5.29 shows an example of the average power delay profile when the transmitter is located on the front of the body and the receiver is on the back of the body. This plot is generated by averaging the 49 power delay profiles obtained in the measurement grid used (see Figure 5.5). Several main signal clusters can be identified. First, a cluster of MPCs is observed shortly after transmission. This component is due to diffraction around the body, and was already analyzed in Section 5.4.3.2. Second, a complex group of overlapping MPC clusters is observed, due to reflections off the ground, ceiling, and nearby objects in the room located in front of the body [3, 38, 46, 54].

A comparison of Figure 5.29 with Table 5.7 shows that if the antennas are placed on the same side of the body, then the received energy due to the reflection off nearby scatterers of MPCs is significantly smaller than the energy received due to the propagation of MPCs near the body, and can be ignored. However, if the antennas are placed on different sides of the body, then the total energy received from scattering in the indoor environment becomes very important, while the initial waves diffracted around the body are significantly attenuated.

Representative time-delay signal parameters in different user scenarios are listed in Table 5.8. As a general conclusion, one can observe a similar range of values determined in the two independent measurement sets, which confirms the validity of the results.

The signal clusters were analyzed in two time-delay regions as described in Section 5.4.1, where the excess delay threshold, T_{env} in (5.4), was determined to be 40 and 25 ns in the static and dynamic channel investigations, respectively. A possible explanation for this dual slope time-delay power decay has been given in Section 5.4.1. The noticeable differences in the excess delay threshold and log-normal cluster fading parameters between the static and dynamic measurements can be explained mainly with the different, and environment-dependent, distribution of

Table 5.7 Large-Scale Distributions Measured in the Static and Dynamic BAN Channel Investigations (Section 5.3.2)

	Static Measurements [48]			Dynamic Measurements [47]								
	Office, Average over All Locations (Figure 5.15)			Laboratory			Corridor			Office		
	Front	Side	Back	Right	Left	Head	Right	Left	Head	Right	Left	Head
Position on the body (Figures 5.13 and 5.17)												
Mean wideband power relative to Tx power (dB)	−69	−73	−78	−52	−61	−56	−49	−59	−56	−47	−58	−53
Wideband power fading σ_s (dB)	0.9	3.1	2.5	1.4	1.5	1.8	1.6	1.0	1.3	1.9	1.2	1.8
Cluster power fading σ_c (dB)	3.3	4.1	2.7	4.5	4.0	4.4	3.9	3.6	3.5	3.8	4.1	3.8

Table 5.8 Time-Delay Channel Parameters Measured in the Static and Dynamic BAN Scenarios

	Static Measurements [48]			Dynamic Measurements [47]								
	Office, Average Over All Locations (Figure 5.15)			Laboratory			Corridor			Office		
	Front	Side	Back	Right	Left	Head	Right	Left	Head	Right	Left	Head
Position on the body (Figures 5.13 and 5.17)												
Excess delay threshold Tenv (ns)		40			25							
Intercluster power decay Γ_1 & Γ_2 (dB/ns)	0.15 & 0.52	0.19 & 0.33	0.11 & —	0.21 & 0.18	0.01 & 0.14	0.05 & 0.17	0.34 & 0.18	0.04 & 0.09	0.10 & 0.14	0.46 & 0.25	0.18 & 0.06	0.18 & 0.20
Cluster arrival rate Λ (1/ns)	0.23	0.13	0.21	0.127	0.139					0.300		
Intracluster power decay γ (dB/ns)	2.4	1.5	1.3	1.6	1.9					1.55		

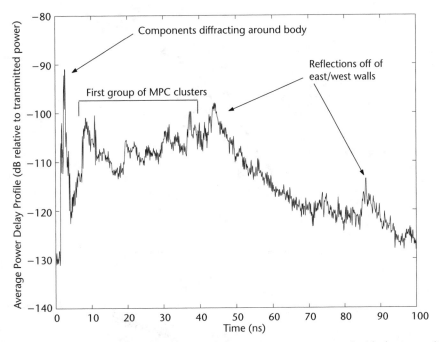

Figure 5.29 Example of average PDP for an indoor environment, measured with the transmit antenna in front of the body, and the receive antenna on the back of the body. (*From:* [48]. © 2006 IEEE. Reprinted with permission.)

the main signal clusters. However, the different procedures used to calculate the average PDP also may have an impact (see Section 5.3.2).

The signal cluster arrival rate and average power decay within the signal clusters show good agreement in the two measurement setups. For simplicity, these parameters are not presented in detail for the dynamic measurements among the antenna positions, due to the very similar range of values of the parameters. Compared to the free-space channel model parameters proposed in IEEE 802.15.4a [56], the determined cluster arrival rates were on average higher in all BAN scenarios. Similar to the explanation of the PAN-MD scenarios in Section 5.4.2, this is due to the presence of the human body in the (static/dynamic) scattering environment surrounding the antennas.

Figure 5.30 shows an example of measured distributions for the time-delay parameters, along with the best-fit normal distribution curves. It can be seen that significant deviations from the normal distribution are to be expected in scenarios with a moving user, and these deviations are generally difficult to reproduce in simulations, unless very specific models are used for different on-body locations. The main reason for this result is believed to be the set of directional antennas used in combination with the movement of the user [47].

5.4.3.4 Channel Modeling

This section summarizes the main radio channel model for the BAN scenarios as proposed in [45, 48, 56], extended with the large-scale time and space domain decorrelation properties proposed in [47].

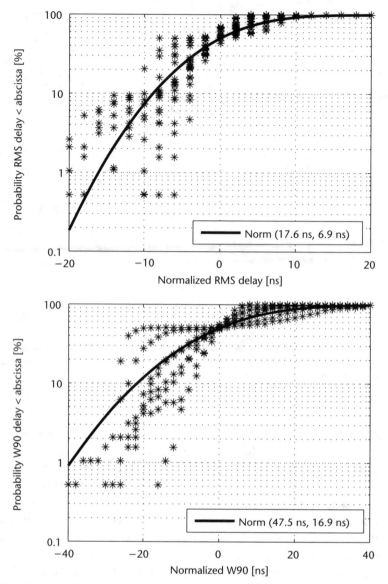

Figure 5.30 Example of measured distribution of the RMS delay spread and 90% energy delay window for all on-body positions with the user walking in the laboratory environment. The fitted normal distributions are given for comparison. (*From:* [47].)

The first cluster of components due to diffraction of the pulse around the body can be modeled as described in Section 5.4.3.1. Subsequent multipath components can be modeled in the same manner as other indoor wireless channels using the parameters described in Section 5.4.3.2. A very common approach that has been adopted by the IEEE UWB standardization committees is the modified SV model [55, 56]. This model uses a double Poisson process to describe the arrival times of multipath components, and assumes that the energy of clusters and rays decay exponentially. The amplitude of each multipath component in the impulse response is generated according to empirically derived correlated log-normal variables.

The proposal for the indoor UWB BAN propagation channel consists of the following main features:

- A signal cluster of components diffracting around the body;
- Deterministic clusters of components due to the geometry of the room;
- Random clusters of components due to reflections off nearby scatterers in the room;
- Large-scale fading of the received energy depending on the position and movement of the body in the room;
- Small-scale fading due to small changes in the position of the body.

Based on the results presented in Section 5.4.3, we build the propagation channel model using a high path loss exponent to describe distance-dependent on-body attenuation, and a discrete time-delay impulse response to describe the scattering phenomena. Similar to the PAN mobile-to-mobile channel results presented in Section 5.4.2, a large-scale approximate decorrelation time of 0.5 seconds and distance of 0.25m was determined in all dynamic BAN investigations with walking users. Therefore, a correlated time and space domain channel realization modeling is proposed, in order to more realistically reflect the behavior of the BAN channel over larger time scales in terms of wideband power and average signal clustering [47].

In a given radio environment (room), there can be important signal clusters that always arrive at the same time and have larger magnitudes than expected, compared to the usual exponential decay trends. A more accurate model of the APDP can be obtained if, in addition to the more random clusters, we add a few deterministic components to account for the geometry of the particular room and the orientation of the on-body devices [48]. This quasideterministic method was used in the channel model derived from the static experiments described in Section 5.3.2.1. However, the model proposal derived from the dynamic investigations described in Section 5.3.2.2 simplifies this modeling by assuming purely random cluster arrival time and a dual exponential decay trend, as in the modified SV models.

For testing purposes, 10,000 channel realizations have been randomly generated and compared with the measured channel impulse responses, in terms of the following important communication channel metrics: RMS signal delay spread (τ_{rms}), 90% energy delay window length ($W90$), and/or a number of significant MPCs (N_p). The RMS delay spread is a good measure of multipath spread over excess delay. The 90% energy delay window parameter can be used to optimize the simple energy detector-like receivers. The number of significant MPCs gives an indication of the number of RAKE fingers required to extract most of the channel energy, and significantly influences the performance of many UWB receivers.

Table 5.9 compares the time-delay parameter statistics from both the measurements and the channel model for the propagation channel around the body without the effects of the environment. The average ($<>$) and standard deviation (STD) values are listed. Good agreement has been found. As expected, the mean RMS delay spread is shorter on the front of the body (0.3 ns) compared with the back of the body (0.5 ns). Similarly, the average number of significant multipath components ranges from 1.8 to 3.0.

Table 5.9 Comparison of Channel Parameters Extracted from Measurements and Simulations for the Propagation Channel Around the Body Without the Effects of the Environment

		Static Measurements [48] Office, Average over All Locations (Figure 5.15)		
		Front	Side	Back
RMS delay spread τ_{rms} (ns)<>/STD	Measured	0.3/0.1	0.4/0.1	0.5/0.1
	Simulated	0.3/0.1	0.4/0.1	0.5/0.1
90% energy window W90 (ns) <>/STD	Measured	1.0/0.3	1.2/0.5	1.4/0.3
	Simulated	1.0/0.3	1.2/0.4	1.4/0.4
Number of significant MPCs N_p (-) <>/STD	Measured	1.9/0.7	2.3/0.8	2.8/0.8
	Simulated	1.9/0.6	2.2/0.8	2.9/0.8

Figures 5.31 and 5.32 compare the cumulative density functions of the τ_{rms} and N_p parameters for both the measurements and the channel model, including the complete propagation on-body channel and the effects of the environment, using the semideterministic modeling approach.

Table 5.10 compares the statistics of the time-delay parameters obtained from the BAN measurements and simulations for the propagation channel around the body including the effects of the environment. The simulations for the channel model extracted from the dynamic measurements were performed without considering different positions on the body. The average (<>) and standard deviation (STD) values are listed, although a simple normal distribution was generally not the best fit, as exemplified in Figure 5.30.

Based on these results, we can conclude that the BAN models are able to capture, up to a certain degree, the different propagation effects, due to the on-body positions and the interaction with the static or dynamic radio environment.

5.4.4 Concluding Remarks

Based on the extensive PAN and BAN radio channel investigation results presented in Section 5.3, channel characterization and modeling for two typical indoor UWB scenarios has been described in this section.

The first model is suited for PAN mobile-to-mobile (body-to-body) scenarios with handheld devices, and represents an enhancement of the existing UWB PAN channel models proposed in the literature. It includes the effects of antenna and device handing, user proximity, and user movement in a given radio environment.

The second model presented is for BAN scenarios with on-body devices. This semideterministic model can be used for simulation of the body-induced propagation channel, as well as for the simulation of the total propagation channel (i.e., including the radio environment of the user). Several modeling extensions are proposed to account for the user dynamics in terms of arm movements and/or movement through the environment (room).

The proposed models have been tested, based on the statistics of three time-delay parameters: RMS delay spread, 90% energy delay window, and number of

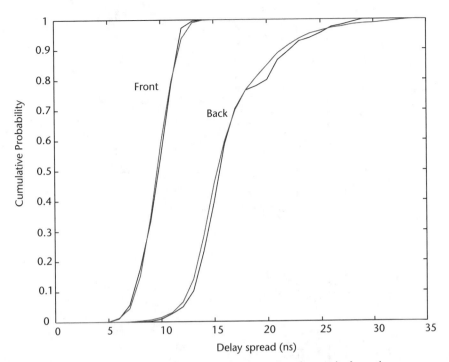

Figure 5.31 Distribution of the RMS delay spread, comparing the results from the measurements (dashed line) and from the model (solid line), for different receiver positions on the body. The signal paths reflecting off the surrounding environment are considered, using the semideterministic modeling approach.

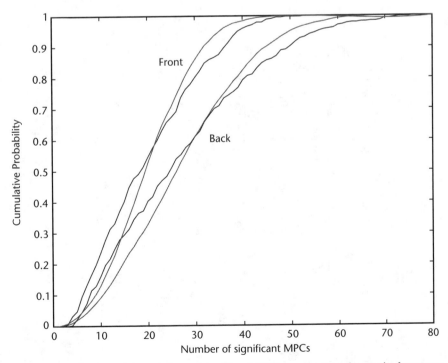

Figure 5.32 Distribution of the number of significant MPCs, comparing the results from measurements (dashed line) and from the model (solid line), for different receiver positions on the body. The signal paths reflecting off the surrounding environment are considered, using the semideterministic modeling approach.

Table 5.10 Comparison of Channel Parameters Extracted from Measurements and Simulations for the Propagation Channel Around the Body Including the Effects of the Environment

Position on the body (Figures 5.13 and 5.17)		Static Measurements [48] Office, Average over All Locations (Figure 5.15)			Dynamic Measurements [47] Laboratory			Corridor			Office		
		Front	Side	Back	Right	Left	Head	Right	Left	Head	Right	Left	Head
RMS delay spread τ_{rms} (ns) <>/STD	Measured	9.6/1.6	14.5/2.0	16.5/3.7	13.6/5.3	22.3/5.8	18/6.4	10.7/5.0	21/12	17.6/6.2	7.6/2.3	16/7	15.1/5.5
	Simulated	9.8/1.6	14.3/3.2	16.2/4.0		17/6		17/6				11/5	
90% energy window W90 [ns] <>/STD	Measured	23/5	35/5	44/10	30/16	68/19	50/20	21/12	63/25	51/23	16/6	60/31	45/19
	Simulated	29/4	34/8	46/12		46/19			45/18			27/14	
Number of significant MPCs N_p [-] /STD	Measured	19/11	21/13	26/16	N/A			N/A			N/A		
	Simulated	20/9	20/10	25/13	N/A			N/A			N/A		

significant multipath components within −10 dB of the largest component in each channel profile. Generally, good agreement has been found with the measurement results. Additionally, the comparison between two independent sets of channel investigations showed agreement for most of the parameters analyzed, but significant differences also have been found. These differences highlight the variability of the extracted channel parameters, not only due to indoor radio propagation environments and on-body positions, but also due to the experimental procedures used.

References

[1]   Gerrits, J., et al., "Design and Performance Analysis of UWB Communication System for Low Data Rate WPAN Applications," *Power Aware Communications for Wireless Optimised Personal Area Network (PACWOMAN)*, IST-2001-34157, Deliverable D 4.1.1, October 2003.

[2]   European IST MAGNET, "Candidate Air Interfaces and Enhancements," IST-507102, My personal Adaptive Global NET (MAGNET), Deliverable D.3.2.2a, October 2004.

[3]   Fort, A., et al., "An Ultra Wideband Body Area Propagation Channel Model: From Statistics to Implementation," *IEEE Trans. on Microwave Theory and Techniques*, Vol. 54, No. 4, Pt. 2, April 2006, pp. 1820–1826.

[4]   Schantz, H. G., and L. Fullerton, "The Diamond Dipole: A Gaussian Impulse Antenna," *IEEE Antennas and Propagation Society Symp.*, July 8–13, 2001.

[5]   Oppermann, I., M. Hamalainen, and J. Iinatti, (eds.), *UWB: Theory and Applications*, New York: John Wiley & Sons, 2004.

[6]   Wu, X. H., and Z. N. Chen, "Comparison of Planar Dipoles in UWB Applications," *IEEE Trans. on Antennas and Propagation*, Vol. 53, No. 6, June 2005.

[7]   Alomainy, A., et al., "Comparison Between Two Different Antennas for UWB On-Body Propagation Measurements," *IEEE Antennas and Wireless Propagation Letters*, Vol. 4, 2005.

[8]   Paulsen, L., et al., "Recent Investigations on the Volcano Smoke Antenna," *IEEE Antennas and Propagation Society Int. Symp.*, Vol. 3, June 22–27, 2003, pp. 845–848.

[9]   Powell, J., and A. P. Chandrakasan, "Differential and Single Ended Elliptical Antennas for 3.1–10.6 GHz Ultra Wideband Communication," *IEEE Antennas and Propagation Society Int. Symp.*, June 2004.

[10] Yazdandoost, K. Y., and R. Kohno, "Design and Analysis of an Antenna for Ultra-Wideband System," *14th IST Mobile and Wireless Communications Summit*, Dresden, Germany, June 19–23, 2005.

[11] Behdad, N., and K. Sarabandi, "A Compact Antenna for Ultrawide-Band Applications," *IEEE Trans. on Antennas and Propagation*, Vol. 53, No. 7, July 2005.

[12] Klemm, M., and G. Troester, "Textile UWB Antennas for Wireless Body Area Networks," *IEEE Trans. on Antennas and Propagation*, September 2005.

[13] Lamensdorf, D., and L. Susman, "Baseband-Pulse-Antenna Techniques," *IEEE Antennas and Propagation Magazine*, Vol. 36, No. 1, February 1994.

[14] McLean, J. S., H. Foltz, and R. Sutton, "Pattern Descriptors for UWB Antennas," *IEEE Trans. on Antennas and Propagation*, Vol. 53, No. 1, January 2005.

[15] Klemm, M., and G. Tröster, "Characterisation of an Aperture-Stacked Patch Antenna for Ultra-Wideband Wearable Radio Systems," *15th Int. Conference on Microwaves and Wireless Communications*, Warsaw, Poland, May 17–19, 2004.

[16] Klemm, M., and G. Tröster, "Small Patch Antennas for Ultra-Wideband Wireless Body Area Networks," *Euro Electromagnetics (EUROEM 2004) Conference*, Magdeburg, Germany, July 2004, pp. 12–16.

[17] Cai, A., T. S. P. See, and Z. N. Chen, "Study of Human Head Effects on UWB Antenna," *IEEE Int. Workshop on Antenna Technology: Small Antennas and Novel Metamaterials, 2005 (IWAT 2005)*, March 7–9, 2005, pp. 310–313.

[18] Alomainy, A., et al., "UWB On-Body Radio Propagation and System Modelling for Wireless Body-Centric Networks," *IEE Proc. Communications, Special Issue on UWB Theory, Technology and Applications*, Vol. 153, No. 1, February 2006, pp. 107–114.

[19] Klemm, M., et al., "Novel Small-Size Directional Antenna for UWB WBAN/WPAN Applications," *IEEE Trans. on Antennas and Propagation*, Vol. 53, No. 12, December 2005, pp. 3884–3896.

[20] Saitou, A., et al., "Practical Realization of Self Complementary Broadband Antenna on Low-Loss Resin Substrate for UWB Applications," *2004 IEEE MTT-S Int. Microwave Symp.*, Fort Worth, TX, June 2004.

[21] Suh, S. -Y., W. L. Stutzman, and W. A. Davis, "A New Ultrawideband Printed Monopole Antenna: The Planar Inverted Cone Antenna (PICA)," *IEEE Trans. on Antennas and Propagation*, Vol. 52, No. 5, May 2004.

[22] Skycross, Inc., Antenna Products, "3.1–10 GHz Ultra-Wideband Antenna for Commercial UWB Applications," January 2003, http://www.skycross.com.

[23] Shih, T.-Y., C.-L. Li, and C.-S. Lai, "Design of an UWB Fully Planar Quasi-Elliptic Monopole Antenna," *2004 Int. Conference on Electromagnetic Applications and Compatibility (ICEMAC 2004)*, Taipei, Taiwan, October 14–16, 2004.

[24] Targonski, S. D., R. B. Waterhouse, and D. M. Pozar, "Design of Wide-Band Aperture-Stacked Patch Microstrip Antennas," *IEEE Trans. on Antennas and Propagation*, Vol. 46, No. 9, September 1998.

[25] Qing, X., M. Y. W. Chia, and X. Wu, "Wide-Slot Antenna for UWB Applications," *IEEE Antennas and Propagation Society Int. Symp.*, Vol. 1, June 22–27, 2003, pp. 834–837.

[26] Toftgard, J., S. N. Hornsleth, and J. B. Andersen, "Effects on Portable Antennas of the Presence of a Person," *IEEE Trans. on Antennas and Propagation*, Vol. 41, No. 6, June 1993, pp. 739–746.

[27] Jensen, M. A., and Y. Rahmat-Samii, "EM Interaction of Handset Antennas and a Human in Personal Communications," *IEEE Proc.*, Vol. 83, No, 1, January 1995, pp. 7–17.

[28] Okoniewski, M., and M. A. Stuchly, "A Study of the Handset Antenna and Human Body Interaction," *IEEE Trans. on Microwave Theory and Techniques*, Vol. 44, No, 10, October 1996, pp. 1855–1864.

[29] Pedersen, G. F., K. Olesen, and S. L. Larsen, "Body Loss for Handheld Phones," *1999 49th IEEE Vehicular Technology Conference*, Vol. 2, May 16–20, 1999, pp. 1580–1584.

[30] Hirata, A., T. Fujino, and T. Shiozawa, "SAR and Temperature Increase Induced in the Human Body Due to Body-Mounted Antennas," *2004 IEEE Antennas and Propagation Society Symp.*, Vol. 2, June 20–25, 2004, pp. 1851–1854.

[31] Christ, A., A. Klingenboeck, and N. Kuster, "Energy Absorption in Layered Biological Tissue and Its Consequences on the Compliance Testing of Body-Mounted Wireless Devices," *Progress in Electromagnetics Research Symp. 2005*, Hangzhou, China, August 23–26, 2005.

[32] Balanis, C. A., *Antenna Theory: Analysis and Design*, 3rd ed., New York: Wiley-Interscience, 2005.

[33] Nielsen, J. Ø., and G. F. Pedersen, "Mobile Handset Performance Evaluation Using Spherical Measurements," *Proc. 2002 IEEE 56th Vehicular Technology Conference, 2002, VTC 2002-Fall*, Vol. 1, September 24–28, 2002, pp. 289–293.

[34] Laitinen, T. A., et al., "Spherical Measurement System for Determination of Complex Radiation Patterns of Mobile Terminals," *Electronics Letters*, Vol. 40, No. 22, October 28, 2004, pp. 1392–1394.

[35] Soergel, W., F. Pivit, and W. Wiesbeck, "Comparison of Frequency Domain and Time Domain Measurement Procedures for Ultra-Wideband Antennas," *Proc. 25th Annual*

Meeting and Symp. of the Antenna and Measurement Techniques Association (AMTA '03), Irvine, CA, October 2003, pp. 72–76.

[36] Kotterman, W. A. Th., et al., "Cable-Less Measurement Setup for Wireless Handheld Terminals," *Proc. Symp. on Personal, Indoor and Mobile Radio Communications (PIMRC),* Vol. 1, October 2001, pp. 112–116.

[37] Hertel, T. W., "Cable-Current Effects of Miniature UWB Antennas," *2005 IEEE Antennas and Propagation Symp.,* Washington, D.C., July 3–8, 2005.

[38] Kovács, I. Z., P. C. F. Eggers, and G. F. Pedersen, "Body-Area Networks," *UWB: Theory and Applications,* I. Oppermann, M. Hamalainen, and J. Iinatti, (eds.), New York: John Wiley & Sons, 2004.

[39] Icheln, C., J. Krogerus, and P. Vainikainen, "Use of Balun Chokes in Small-Antenna Radiation Measurements," *IEEE Trans. on Instrumentation and Measurement,* Vol. 53, No. 2, April 2004, pp. 498–506.

[40] Manteghi, M., and Y. Rahmat-Samii, "A Novel UWB Feeding Mechanism for the TEM Horn Antenna, Reflector IRA, and the Vivaldi Antenna," *IEEE Antennas and Propagation Magazine,* Vol. 46, No. 5, October 2004, pp. 81–87.

[41] Kovács, I. Z., P. C. F. Eggers, and K. Olesen, "UWB Radio Propagation Investigations in Body Area Network Scenarios," *IEEE Proc. 8th Int. Symp. on Spread Spectrum Techniques and Applications (ISSSTA),* Sydney, September 2004.

[42] Klemm, M., and G. Troster, "Integration of Electrically Small UWB Antennas for Body-Worn Sensor Applications," *2005 IEE Wideband and Multiband Antennas and Arrays,* 2005, Birmingham, U.K., pp. 141–146.

[43] Manteuffel, D., "FDTD Characterization of UWB Antennas for Home-Entertainment Equipment with Special Emphasis on the Specific Integration Scenario," *IEEE Int. Workshop on Antenna Technology: Small Antennas and Novel Metamaterials, 2005 (IWAT 2005),* March 7–9, 2005, pp. 351–354.

[44] United States National Library of Medicine, National Institutes of Health, Health & Human Services, "The Visible Human Project," http://www.nlm.nih.gov/research/visible/visible_human.html, July 2005.

[45] Fort, A., et al., "Ultra Wide-Band Body Area Channel Model," *Int. Conference on Communication (ICC),* May 2005.

[46] Fort, A., et al., "Characterization of the Ultra Wideband Body Area Propagation Channel," *Int. Conference on Ultra Wideband (ICU),* September 2005.

[47] European IST MAGNET, "PAN Channel Characterisation (Part I & II)," IST-507102, My personal Adaptive Global Net, Deliverable D.3.1.2a&b, October 2004 and June 2005.

[48] Fort, A., et al., "Ultra Wideband Channel Model for Communication Around the Human Body," *IEEE J. on Selected Areas in Communications (JSAC),* Vol. 24, No. 4, April 2006, pp. 927–933.

[49] Kovács, I. Z., et al., "UWB Radio Channel Model for Short-Range Mobile-to-Mobile Communication Scenarios," *Wireless Personal Multimedia Communication Conference,* September 2005.

[50] Cassioli, D., M. Z. Win, and A. F. Molisch, "The Ultra-Wide Bandwidth Indoor Channel: From Statistical Model to Simulations," *IEEE J. on Selected Areas of Communications,* Vol. 20, No. 6, 2002.

[51] Zetik, R., R. Thomä, and J. Sachs, "Ultra-Wideband Real-Time Channel Sounder Design and Application," *URSI Int. Symp. on Electromagnetic Theory,* Pisa, Italy, May 2004.

[52] Kovács, I. Z., et al., "Enhanced UWB Radio Channel Model for Short-Range Communication Scenarios Including User Dynamics," *14th IST Mobile and Wireless Communication Summit,* June 2005.

[53] Ciccognami, W., A. Durantini, and D. Cassioli, "Time Domain Propagation Measurements of the UWB Indoor Channel Using PN-Sequence in the FCC-Compliant Band 3.6–6 GHz," *IEEE Trans. on Antennas and Propagation,* Vol. 53, No. 4, April 2005, pp. 1542–1549.

[54] Kovács, I. Z., and P. C. F. Eggers, "UWB Radio Channel Characterisation for Portable User Terminal Scenarios," *Future Adaptive Communication Environment (FACE) Research Project*, Deliverable D4.2, Center For PersonKommunikation Center for TeleInFrastruktur, Aalborg University, Denmark, March 2004, http://www.cpk.auc.dk/FACE.

[55] Foerster, J., (ed.), "IEEE 802.15.3a Channel Modelling Subcommittee—Report Final," *IEEE P802.15 Working Group for WPAN*, November 2002.

[56] Molisch, A. F., (ed.), "IEEE 802.15.4a Channel Model Subgroup Final Report," *IEEE P802.15 Study Group for WPAN*, September 2004.

[57] Molisch, A., et al., (eds.), "UWB Communication Systems—A Comprehensive Overview," *EURASIP Book Series in Signal Processing and Communications*, Hindawi Publishing, ISBN 977-5945-10-0, 2005.

[58] Zasowski, T., et al., "UWB for Noninvasive Wireless Body Area Networks: Channel Measurements and Results," *Proc. 2003 IEEE Conference on Ultra Wideband Systems and Technologies (UWBST)*, October 2003, pp. 285–289.

[59] Saleh, A., and R. A. Valenzuela, "A Statistical Model for Indoor Multipath Propagation," *IEEE J. on Selected Areas of Communications*, Vol. 5, February 1987, pp. 128–137.

[60] Hashemi, H., "The Indoor Radio Propagation Channel," *IEEE Proc.*, Vol. 81, No. 7, 1993, pp. 943–968.

[61] Álvarez, Á., et al., "New Channel Impulse Response Model for UWB Indoor System Simulations," *IEEE Vehicular Technology Conference*, Jeju, Korea, April 21–24, 2003.

[62] Kunisch, J., and J. Pamp, "An Ultra-Wideband Space-Variant Multipath Indoor Radio Channel Model," *IEEE Conference on Ultra Wideband Systems and Technologies (UWBST)*, November 2003.

CHAPTER 6

Wearable Antennas: Advances in Design, Characterization, and Application

Pekka Salonen and Yahya Rahmat-Samii

6.1 Background

The evolution of antenna technology for man-machine interface has taken quantum leaps in utilizing textile materials as antenna substrates. In the future, this will allow complete freedom to design body-worn antenna systems embedded in so-called "smart clothes." Smart clothes soon may find their place in our everyday living. They will emerge in various sports outfits, emergency workers' outfits, military, medical, and space applications, and so forth. The ability to establish wireless communication is one mandatory requirement for smart clothes. In addition, wireless data communication via smart clothes offers a host of entertainment possibilities [1].

The earliest demand for wearable antennas was set by military applications, where whip antennas carried by the person were easy targets to identify the radio in a squad. The identification of the radio operator can be hindered if the antenna is incorporated into the uniform. In addition, this incorporation of the antenna into the uniform has the added benefit of eliminating clumsy devices that can tangle in trees, foliage, and low-hanging obstructions. This has led to the development of an RF helmet antenna [2, 3] and an RF vest antenna [4–6]. Both these antennas operate over a wide frequency band. The helmet antenna is constructed of conductive cloth that has polyester interwoven with nickel and copper fibers [2]. The vest antenna is constructed from canvas and FLECTRON, which is a mixture of copper interwoven with polyester to form a highly conductive cloth [4]. The advantages of these antennas are that they are lightweight and inexpensive, and require very low maintenance. There is no visual signature provided, which allows the radio operator to stay indistinguishable from any other soldier. The wearable antennas require no complicated setup, are not easily damaged from foliage or obstacles, and permit all-weather operation.

Extensive worldwide research has been carried out on new medical imaging methods. An ideal medical screening tool is noninvasive, is simple to use, has a low health risk, and is sensitive to tumors and specific malignancies. It also needs to be cost-effective and widely available, minimize discomfort, allow easy interpretation, and provide consistent results. One of the future imaging methods will employ nonionizing electromagnetic waves to image the human body, as a means of detecting cancer cells as early as possible. Promising results of microwave imaging have been reported in the field of early breast cancer detection [7–10]. This is due to the fact that breast tumors have electrical properties at microwave frequencies that are significantly different from the properties of healthy breast tissues. The breast is illuminated with microwaves, and the transmitted or reflected microwave signals are measured. According to [7], current microwave imaging methods for breast cancer employ a set of monopole antennas. However, research is underway to develop better antennas for microwave imaging, and textile antennas that are comfortable against the skin may be applicable in these areas [7, 8].

Proceedings of the IEEE International Symposium on Wearable Computers show that wearable computers have a history dating back to 1993 [11]. They can be considered as ancestors to smart clothes [12]. Smart clothes are intelligent wearable systems, which can be used in sportswear, emergency workers' outfits, astronauts' suits, and in medical, military, and entertainment applications. The intelligence in clothing is composed of intelligent textile materials, consisting of electrical and nonelectrical components. The traditional function of clothes can be extended to comprise new tasks, such as survival assistance in demanding conditions. This goal can be achieved by embedding or integrating additional components into clothing, which will provide new functions, such as calling for assistance or monitoring medical conditions. Recently, smart clothes were introduced for the Arctic environment [12–14]. Smart clothes are commercially available in the form of wearable phones (e.g., wearaphones), as shown in Figure 6.1 [15, 16].

The first public research report on wearable antennas dates back to 1999, when a dual-band planar antenna was designed for wearable and ubiquitous equipment

Figure 6.1　Example of the commercial smart clothing. (Courtesy of Reima Smart Clothing, Finland.)

[17]. Since then, the research on wearable antennas has received significant interest among university and industry researchers [2–6, 18–32]. The wearable antenna can be defined as an antenna that is designed and meant to be a part of clothing [17].

In this chapter, attention is focused on miniaturized wearable antennas made from various textile materials. Critical design parameters of wearable antennas for various applications are discussed. Examples for cellular phone, Global Positioning System (GPS), and WLAN applications are presented. This chapter also addresses new methods for improving wearable antenna performance. Examples will cover wearable dual-band and electromagnetic bandgap (EBG) antennas. The effect of the vicinity of the human body on wearable textile antenna performance is discussed in detail. The terms textile and fabric antenna are used interchangeably here. However, in most cases, textile antenna refers to an antenna type, whereas the term fabric relates the antenna substrate material to a particular textile material, such as fleece fabric.

6.2 Wearable Antennas: Critical Design Issues

For flexible antennas, textile materials form interesting substrates, because fabric antennas can be easily integrated into clothes. Textile materials generally have a very low dielectric constant, which reduces the surface wave losses and improves the impedance bandwidth of the antenna. In comparison with the high dielectric substrates, textile antennas are physically larger. The variation of the dimensions due to stretching and compression are typical for fabrics, and this has have a strong influence on the electromagnetic characteristics of the antenna. The changes in the resonant length of the antenna detune its frequency band. The substrate thickness changes the resonant frequency as well as the input impedance bandwidth, which can be characterized by measuring the return loss of the antenna using a network analyzer.

Recently, textile antennas have been presented at international conferences [22, 25–32]. These studies have demonstrated that textile materials are suitable for microstrip antenna substrates. In [24], the first fleece fabric GPS antenna developed was discussed, and in [26–28], WLAN antennas were presented. However, GPS-operated systems require circularly polarized antennas, whereas WLAN needs linearly polarized antennas. In general, circularly polarized antennas need not have mathematically perfect circular polarization in order to function, but elastic antenna materials complicate the design process. Therefore, it is challenging to design a fabric antenna that has stable electrical characteristics.

The wearable antenna requirements are application-specific; common requirements for many applications are: (1) lightweight, (2) inexpensive, (3) low-maintenance, (4) no setup requirements, and (5) able to withstand damage from obstacles (robust). In order to present the key components of wearable antennas, we will address several major design features of these antennas in the following sections. The important design features include: (1) selection of textile material; (2) antenna performance that depends on material properties, wrinkle, and so forth; (3) the influence of ground plane size; (4) effects of material conductivity; (5) performance enhancement using electromagnetic bandgap structures; (6) effects of the human body on the overall wearable antenna performance, including optimal positioning of the antenna on the human body; and (7) the SAR. Figure 6.2 summarizes the key features related to design process for wearable antennas. To address these issues, we use various application scenarios, such as GPS, WLAN, and so forth. It must be

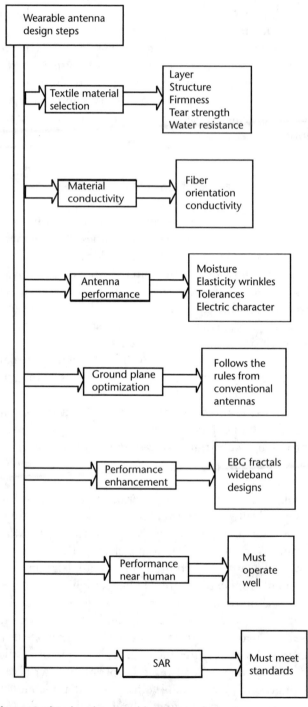

Figure 6.2   Key features related to the wearable antenna design process.

pointed out that when comparing input-matching plots, the scale may vary in order to highlight the overall performance. The radiation patterns are generally plotted with respect to dBi.

6.3 Textile Materials

Textile materials that are used as substrates for antennas can be divided into two main categories—natural and synthetic fibers. Synthetic fibers are polymers obtained from their molecular structures. The subsets of polymers have the prefix "poly-," as shown in Figure 6.3. The names of the fibers are generally trademarks of companies, and they are classified based on their typical radical. Thus, polyesters (for example) can have many different molecular formulas and many trademarks [33–35]. One commonly known polymer is polytetrafluoroethylene (PTFE), which is better known as Teflon.

Polymers are extremely long molecules that consist of a repeating molecular structure. Very thin polymer fibers can be wound to yarn, woven, or knitted. Dacron is a widely used polyester in textile materials. Five different synthetic fabric materials used in this study are fleece, upholstery fabric, vellux, synthetic felt, and Cordura. Fleece is a synthetic woven polyester fabric, which is felted from both sides and is very soft and comfortable against the skin. Because polyester fibers are hydrophilic, they repel rather than absorb water. Thus, polyester fleece dries very rapidly. In this study, the fleece fabric thickness is 4 mm. Fleece fabric is commonly used for sportswear and leisurewear.

Upholstery fabric is composed of a mixture of polyester and polyacryl. It is woven, with three fabric layers bound together by bounding wefts. It is quite thin,

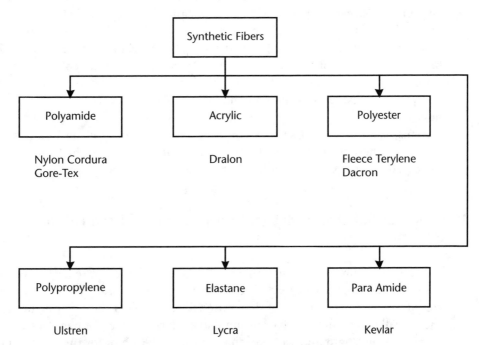

Figure 6.3 Classification of synthetic fibers with example fabrics.

1.1 mm, but its firmness makes this material an interesting choice for antenna designs. It does not bend as easily as does soft fleece.

Vellux has hairy synthetic fabric in the center that is covered from both sides with a thin layer of foam plastic, with the hair penetrating into the foam. As a result, it is a soft collapsible fabric, with good insulating properties. Vellux is somewhat elastic in both directions. The fabric thickness is 5 mm.

Synthetic felt resembles felt made out of wool, but the fabric is not woven or knitted. It is about 4 mm thick, and the fibers are looser on the surface than at the center. It can be easily deformed, and does not revert to its initial shape and dimensions.

Cordura is a trademark of DuPont, and the Cordura used in our study is made by Delcotex, in Germany [36]. The name of the fabric is Delinova 200, which is woven from Cordura fibers. Cordura is polyamide fiber, well-known as nylon. Delinova 200 has a fluorocarbon impregnation and is coated with polyurethane. It weighs about 370 g/m^3, is very hard, and has a constant thickness of 0.5 mm. It has very high tensile and tear strengths in the warps and wefts directions [36].

When considering the elastic properties of fabrics, the texture and the material are decisive. Most fabrics are inelastic, and real elastic fabrics are mostly knitted from elastic fibers or yarn. Since elastic fabrics are used as antenna substrates, problems may occur. When stretched, the permittivity and thickness of the fabric change, which changes the antenna resonant frequency. Stretching makes it difficult to attach the metallic layers onto the fabric, and just a few detached sections significantly changes the properties of the antenna. Delinova 200–Cordura is an interesting choice of fabric for antennas, not only because of its strength and constant thickness, but also because of its almost perfect water resistance. It resists a very high water column, more than 1m [36]. Water has a much higher dielectric constant than does the fabric. When a fabric antenna absorbs water, the moisture changes the antenna performance parameters dramatically. The higher dielectric constant of water dominates the antenna performance by reducing the resonant frequency. Since fabric antennas are used near the skin, the aspect of the wetness of the fabric becomes more important. The other fabrics studied here, except Cordura, absorb water easily. Therefore, the changes in performance parameters become more obvious with increasing moisture content. The only problem with Cordura is its thickness (1.1 mm) when designing wideband antennas. However, for narrowband systems, such as GPS, Cordura could be ideal. In addition, in multilayer textile antennas where the outermost layer is made out of Cordura, the bandwidth can be increased and the water resistant characteristics can be retained.

6.4　Effects of Substrate Materials: An Example of the Fabric GPS Antenna

GPS antenna dimensions were calculated using FDTD software developed at UCLA ARAM Laboratory. Antennas were constructed out of fleece, felt, vellux, and upholstery fabrics. The geometry of the fabric GPS antenna is shown in Figure 6.4. Both the overall length and width of the antenna determine the resonant frequency of the antenna. The length and width of the antenna are 88 mm for the fleece fabric ($\varepsilon_r = $

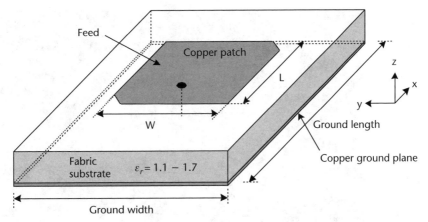

Figure 6.4   GPS antenna geometry residing on a fabric substrate.

1.1). The corners of the patch are cut off, in order to obtain right-hand circular polarization, which is required for GPS receiver antennas. The ground plane dimensions are 130 × 130 mm. Therefore, the ground plane is 16 mm larger in all directions than the radiating patch element itself. The effect of ground plane size is discussed in detail in Section 6.4.1. Conductive parts are made out of copper tape, and the substrate material is conventional fleece fabric. The substrate thickness is 4 mm. The copper tape and the fleece fabric together make the antenna flexible. A coaxial SMA connector was used to feed the antenna, which was soldered onto the copper tape patch and ground plane. The dielectric constant of a piece of the fabric can be found by using the so-called cavity perturbation method. In this method, the resonant frequency of an empty cavity is measured. Then the cavity is filled with the desired dielectric material, and the resulting resonant frequency is measured. The ratio of the two measured resonant frequencies gives the square root of fabric dielectric constant. Here, the sample fabric was not allowed to compress. The measured dielectric constant at 1.575 GHz for the various fabrics is typically between 1.1 and 1.7. Figure 6.5 shows photos of three constructed prototypes made out of different assemblies of upholstery fabric and Cordura fabric.

An antenna can be considered as a load to an RF preamplifier. This load can be characterized in terms of impedance, which is equivalent to the input impedance of the antenna. In order to couple the maximum available power from an amplifier, the input impedance of the antenna must be matched to the output impedance of the amplifier. The values for the input impedance are usually specified as a voltage standing wave ratio (VSWR) when it is less than 2:1. A 2:1 VSWR is equivalent to a 10-dB return loss, which means 10% of the incident power is reflected back to the source (i.e., power amplifier).

The input impedance of a circularly polarized antenna typically has two resonances close to each other. This is due to the fact that the primary and secondary resonant modes of the antenna are determined by the length and width of the patch. These two modes can be combined by properly adjusting the ground plane size and the dimensions of the cut corners, which makes the rectangular patch antenna circularly polarized. This effect was examined with a fleece antenna, and it was seen that the more the corners were cut, the further the minima of the two resonant modes got from each other on the frequency axis. Therefore, right angle triangles with 10-mm

Figure 6.5   Examples of GPS antennas on various textile materials: (a) upholstery fabric, (b) Cordura, and (c) two-layer upholstery fabric. (Courtesy of Mikko Keskilammi.)

legs were cut from the two corners. The minima stayed fairly close, and the antenna was circularly polarized.

The distance between the radiating patch and the ground plane should remain constant in order to maintain the electrical characteristics of the antenna. In this study, all-conductive sheets were made out of a copper tape in order to accurately characterize the effect of textile materials. If the copper tape detaches only from one corner and the space between the metal layers varies, the resonant frequency of the antenna changes. The same effect occurs when the antenna is bent, as shown in Figure 6.6. The changes of electromagnetic characteristics caused by transformations are difficult to predict, since fabrics can take many diverse shapes. This kind of behavior was studied in practice, particularly with fleece and upholstery fabrics. The copper tape was tightly attached to the felt surface of fleece, and care was taken to ensure that the tape did not detach. However, the fleece antenna changed its shape too easily, due to its softness. Bending effects on the radiation pattern are discussed in [26].

The copper tape did not fasten itself properly onto the surface of the more rigid upholstery fabric. A detached edge or corner caused a change in the resonant frequency of the antenna, but when the detached part was fastened back, the original resonant frequency was obtained, as shown in Figure 6.6. This experiment provides a good example of the importance of the proper attachment method.

The thickness of the textile material has a great influence on the antenna bandwidth. Here, vellux, felt, fleece, and upholstery fabric all have nearly the same permittivity ($\varepsilon_r = 1.1$ to 1.2); therefore, the thickness generally determines the bandwidth. Figure 6.7 shows that it is possible to design well-matched input impedances for all the fabric substrates.

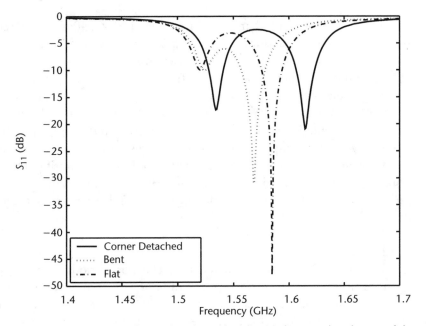

Figure 6.6   Resonant frequency fluctuation due to bending and corner detachment of the antenna.

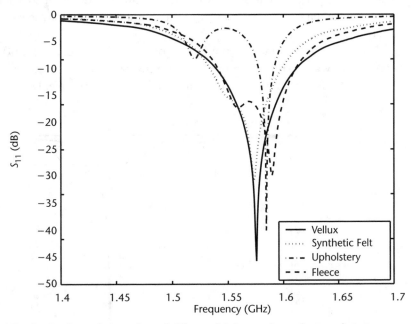

Figure 6.7   Comparison of return loss of different fabrics used as antenna substrates.

6.4.1   Effects of Ground Plane Size Attached to the Fabric Substrate on GPS Antenna Performance

The effect of different ground plane sizes on fabric antennas was examined with upholstery and Cordura fabrics. Two upholstery antennas were built with identical patches on rectangular ground planes. The dimension of the smaller ground plane was 100×100 mm, and that of the larger was 140×140 mm. The return losses of both antennas were measured. The difference in the ground plane size has minimal effect on the resonant frequency, as shown in Figure 6.8.

Antennas were measured in an electromagnetically anechoic chamber with fully automatic measurement software. Polarization measurements were performed using the polarization-pattern method [37]. A linearly polarized standard gain horn antenna was aimed at the antenna under test, rotated through a full 360°, and relative gain measurements were recorded. The resulting pattern gives the axial ratio (AR) but not the polarization sense or the polarization purity. The axial ratio is obtained by first measuring the vertical radiation pattern and then the horizontal radiation pattern. Based on the difference between these patterns, the axial ratio can be calculated.

The main difference between the measured radiation patterns is their gain, which can be seen from Figures 6.9 and 6.10. However, both prototyped antennas have good circular polarization characteristics. The axial ratio is approximately 1 dB.

Due to the unstable geometry, the radiation pattern differs slightly in identical measurements. The measurement arrangement also causes an error in the result, but it is hard to draw a line between these sources of error. Therefore, to evaluate the collective error, the upholstery antenna with a 140×140 mm ground plane was measured eight times in the same position, E-plane (xz-plane). Between the measurements, the antenna was sometimes unfastened, fastened back, or straightened, and the maximal variation from the maximum value of the gain was approximately 2 dB.

A preliminary prototype made out of Cordura—Delinova 200 with circular polarization was designed. Figure 6.11 summarizes measured the radiation patterns with 32-dB offset in plots, which are normalized to a dBi scale. Due to the properties

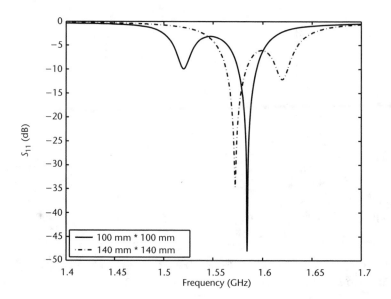

Figure 6.8 Measured return loss with different ground plane sizes.

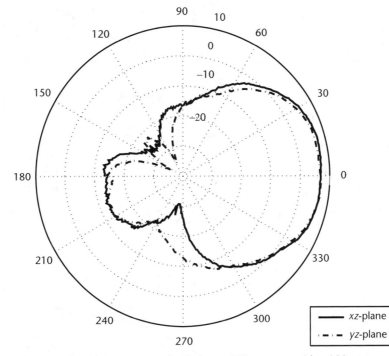

Figure 6.9   Measured radiation pattern of upholstery GPS antenna with a 100-mm square ground plane.

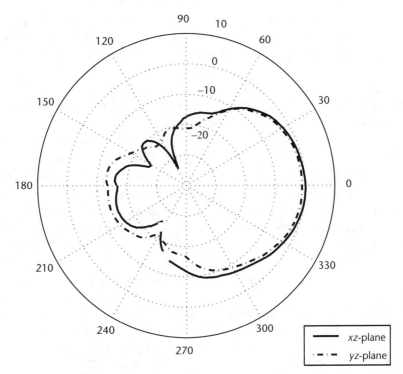

Figure 6.10   Measured radiation pattern of upholstery GPS antenna with a 140-mm square ground plane.

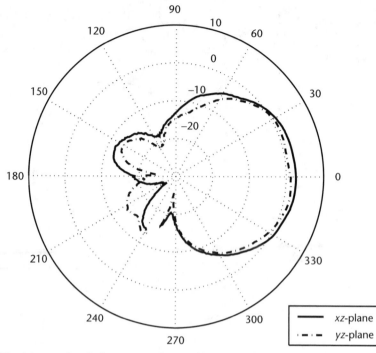

Figure 6.11 Measured radiation pattern for Cordura GPS antenna with a 140-mm square ground plane.

of Cordura presented in the previous section, it does not stretch or contract as do other fabric materials. Therefore, Cordura is more suitable for circularly polarized antennas with stable antenna performance parameters.

6.5 Effect of Various Conductive Materials of Patch Antennas: An Example of WLAN Antenna on Fleece Fabric

The effect of the conductive layer material on textile antenna performance was studied with six different fabricated WLAN antennas. The patch antennas were mounted on the fleece fabric. The thickness of the fleece fabric was 3 mm, and the measured relative permittivity of the substrate was 1.1 at 2.45 GHz.

The conductive surfaces were made out of different materials, of which two were conductive fabrics (knitted copper and Aracon fabric), and different copper tape assemblies were used. These different assemblies were included because we were trying to simulate possible fabric discontinuations. In the first copper tape antenna, the patch was made out of solid homogeneous copper tape, while in the second antenna, the patch was made out of vertically cut pieces of copper tape. In the third antenna, the patch was made out of horizontally cut pieces of copper tape, as shown in Figure 6.12. The horizontally cut pieces of the copper tape were soldered together in one case, and the pieces were glued together in the other case.

The textile antenna dimensions were first adjusted to correspond to the desired center frequency 2.45 GHz, using the method of moments. According to the simulations, the patch length was 56 mm and width was 51 mm. The feed was located

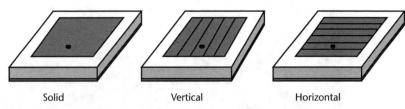

Solid   Vertical   Horizontal

Figure 6.12   Three different copper tape assemblies studied.

19 mm from the patch edge. The simulated return loss is shown in Figure 6.13. Antennas were fabricated based on the simulations, with a 76 × 71 mm ground plane in all cases.

Figure 6.14 shows the fabricated antenna where knitted copper fabric was used in the patch as well as in the ground plane. Antennas are named as follows: antenna A, solid copper tape; antenna B, knitted copper fabric; antenna C, vertically cut copper tape; antenna D, horizontally cut copper tape; antenna E, horizontally cut and soldered copper tape; and antenna F, Aracon fabric. The radiation patterns at 2.45 GHz in both the E-plane and H-plane (*xz*-plane and *yz*-plane, respectively), and the return loss, were measured for all antennas. The return loss was measured using an HP8722D network analyzer. The radiation patterns were measured in an electromagnetically anechoic chamber using an HP8590 spectrum analyzer and a fully automated measuring system. The antenna gain was measured using the gain transfer method. All six antennas were measured under similar conditions, and they were compared in terms of antenna gain, radiation pattern, half-power beamwidth, impedance bandwidth, and impedance matching. The results showed that the conductive material plays an important role in optimal textile antenna design.

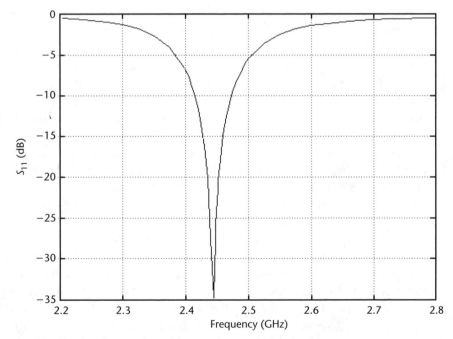

Figure 6.13   Simulated return loss of fleece substrate with knitted copper.

Figure 6.14   Fabricated textile antenna with knitted copper and fleece fabric. (Courtesy of Mikko Keskilammi.)

Figures 6.15 and 6.16 show the measured return loss for all six antennas. Antennas A, C, and E have very similar results. The impedance of these antennas is well-matched at the desired frequency. The impedance bandwidths are similar, as shown in Table 6.1. Antenna B also has a good impedance match, and the impedance bandwidth is wider compared to the antennas made out of copper tape. However, the increase in impedance bandwidth is due to the increased conductor loss. As can be seen, antennas D and F have notably worse performance. Both impedance match and bandwidth have degraded. The results indicate that both conductive tapes and fabrics are possible conductive layer materials for fabric antennas, in terms of impedance matching and bandwidth. However, observing the results obtained from the different configurations, one can see that not all copper tape and fabric configurations are suitable for fabric antennas. If the conductive fabric is used, then it must have good conductivity and be densely knitted. The Aracon fabric has poor results even though it is quite densely knitted. This is because the thread has insufficient conductivity.

The results also show that the conductive material need not be homogeneous in order to obtain good impedance match and bandwidth. Both the solid copper tape and the vertically cut copper tape had very similar results, although the pieces of copper tape were not even soldered together. The conductive elements can have discontinuities as long as they are parallel to the surface current. When the discontinuations are parallel to the surface current, they will not cause reflections of EM fields. The antenna fabricated on horizontally cut copper tape had notably worse performance compared to the other configurations, due to discontinuations perpendicular to the current flow direction. However, when the pieces were soldered together, the results corresponded closely with the results obtained with solid copper tape.

Figure 6.17 shows the radiation patterns of all six antennas. Antenna gains, cross-polarization ratios, and beamwidths are listed in Table 6.2. The antenna gain

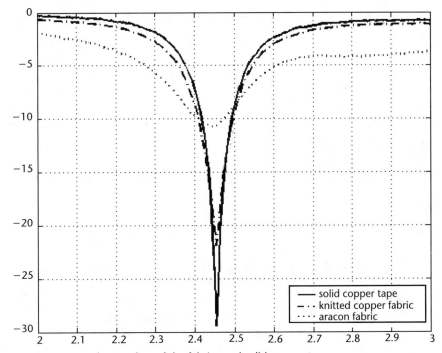

Figure 6.15   Measured return loss of the fabrics and solid copper tape.

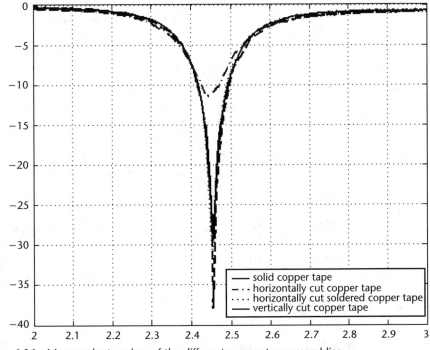

Figure 6.16   Measured return loss of the different copper tape assemblies.

was affected by conductive materials, in the same way as the impedance match and bandwidth are affected. Solid copper tape, vertically cut copper tape, and knitted

Table 6.1 Measured Antenna Parameters with Different Conductive Layer Materials

	Resonant Frequency (MHz)	Input Return Loss (dB)	Bandwidth (MHz) VSWR = 2:1
Solid copper tape	2,455	−29.45	2,417–2,494 → 77
Knitted copper fabric	2,455	−21.88	2,411–2,497 → 86
Vertically cut copper tape	2,455	−37.88	2,416–2,498 → 82
Horizontally cut copper tape	2,445	−11.40	2,427–2,473 → 46
Horizontally cut and soldered copper tape	2,455	−31.76	2,412–2,492 → 80
Conductive fabric	2,450	−10.78	2,411–2,476 → 65

copper fabric, as well as horizontally cut and soldered copper tape, all had similar gain values. The measured gain is approximately 7 dBi. The other two antennas (antennas D and F) have notably lower gain. The discontinuities perpendicular to the current flow decrease the antenna gain more than 2 dB. Horizontal discontinuities (antenna D) corrupt the antenna's performance, since the current flow is interrupted. Discontinuities result in power losses that are likely caused by an increase in ohmic losses and spurious radiation from discontinuities. The poor performance of Aracon fabric results from low fabric conductivity.

The conductive material has an effect on cross-polarization ratio, but does not have a notable effect on the beamwidth. Cross-polarization ratios vary from −12 to −25 dB. The comparison of the cross-polarization ratios suggests that the reason for the low gain and overall weak performance of the Aracon fabric antenna could be partly caused by the increased cross-polarization level. More power is transmitted to undesired directions, resulting in lower gain. The lower gain of the knitted copper fabric could be explained by the decrease in the cross-polarization ratio compared to antennas with copper tape. On the other hand, the weaker performance of the horizontally cut copper tape antenna was related to the greater power losses, and not to the increased cross-polarization radiation.

Comparisons show that the conductive material has very little effect on beamwidth. All of the six antennas had similar values. As shown in Table 6.2, all six antennas have beamwidths around 75° in the E-plane, and beamwidths around 60° in the H-plane.

All antennas have a wide main beam, and because of a small ground plane, a large backlobe radiation. The nulls are approximately 90° from the maximum direction of the main lobe, resulting in a wide main beam. The measured radiation patterns in both the E-plane and H-plane correspond well with each other. In both the E-plane and H-plane, the maximum received power levels were almost the same, with a slight difference on the main beamwidth.

6.6 Dual-Frequency Wearable Antenna Design: An Example of a U-Slot Patch

Wearable systems can exploit different radio systems to communicate between the antennas within the clothing and the outside world. For example, wireless

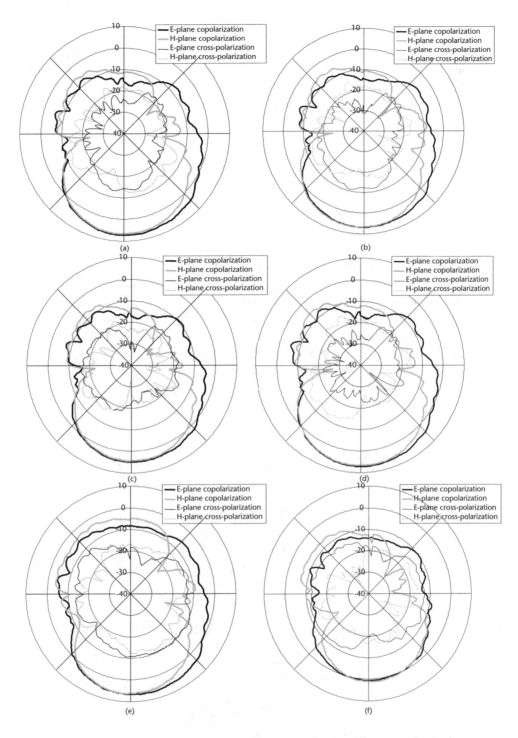

Figure 6.17 Measured radiation patterns (with respect to dBi) for different conductive layer material. The radiation patterns represent the gain values in different directions. In these figures the antenna boresight is downward. (a) Solid copper tape, (b) vertically cut copper tape, (c) horizontally cut copper tape, (d) horizontally cut and soldered copper tape, (e) knitted copper fabric, and (f) Aracon fabric.

Table 6.2   Measured Radiation Parameters with Different Conductive Layer Materials

	Gain (dBi)	X -Polarization Ratio (dB)		Beamwidth (°)	
Solid copper tape	7.35	−20.42 (E)	−18.55 (H)	73 (E)	60 (H)
Knitted copper fabric	6.77	−16.03	−17.74	75	54
Vertically cut copper tape	6.82	−17.73	−18.52	74	61
Horizontally cut copper tape	5.01	−20.83	−17.99	74	62
Horizontally cut and soldered copper tape	7.26	−24.72	−18.90	73	62

short-range links are useful for sending commands from a central processing unit (CPU) to remote-located sensors in the clothing, and for sending data from the sensors back to the CPU for further processing. A long-range wireless communication can be used for location purposes (GPS) or cellular phone applications, such as smart messaging, browsing the Internet, call dialing, and so forth. These aspects set the demand for wearable dual-band antennas.

Figure 6.18 shows the geometry and the dimensions in millimeters of the U-slot patch antenna mounted on the surface of a 3.5-mm-thick fleece fabric. The length and width of the ground plane are 110 and 130 mm, respectively. The measured relative permittivity of the substrate is 1.1. The conducting surfaces are made out of

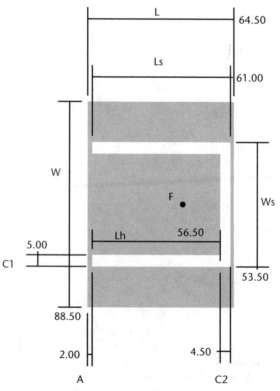

Figure 6.18   Geometry of the prototype U-slot dual-frequency antenna, with dimensions in millimeters.

copper tape with a thickness of 0.075 mm. The dimensions L and L_h are the critical antenna dimensions. The dimension L_h is approximately $\lambda/2$, where λ is determined by the higher resonant frequency. The antenna length L determines the lower resonant frequency. Therefore, L is approximately $\lambda/2$ without the presence of the U-slot. However, the presence of the U-slot lowers the resonant frequency by 5%. In this study, the effect equals approximately 200 MHz. Thus, the L needed to be shortened. The size of the ground plane was selected to be large enough to achieve acceptable performance with dimensions 110 × 130 mm, but the ground plane dimensions were not optimized in this study.

The antenna is fed by a probe feed soldered at point F, as shown in Figure 6.18. The feed is located at (42 mm, 44 mm). The feeding position was optimized using MoM simulations. A change in the feeding position alters the impedance match at both frequencies, while the resonant frequencies remain unchanged.

6.6.1 Experimental Results and Discussions: Fleece and Vellux Fabrics

The wearable dual-band textile antenna shown in Figure 6.18 was prototyped and measured. The photo of the handmade prototype is shown in Figure 6.19. The wrinkles on the antenna copper tape were also present in the measurements. The slot was created in the copper tape, and the conducting elements were mounted on fleece fabric. The ground plane was also made out of copper tape. In order to study the radiation characteristics of the fabricated antenna, the radiation patterns for both frequencies were measured, in both the H-plane and the E-plane.

Figure 6.20 shows the measured and simulated return loss of the constructed prototype. The simulated results were obtained using MoM. The results show good agreement between the measured and simulated return loss, even though the antenna has wrinkles. The impedance is well-matched at the desired frequency in

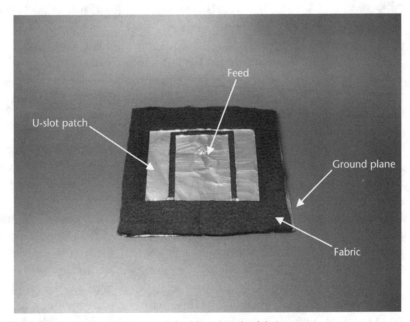

Figure 6.19 Constructed and measured dual-band U-slot fabric antenna.

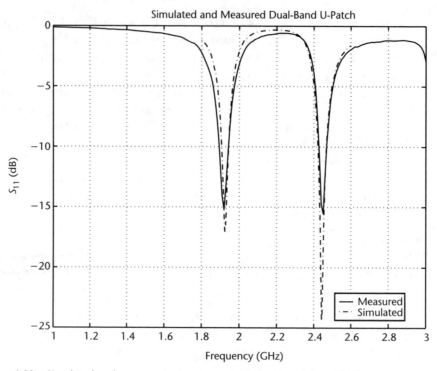

Figure 6.20   Simulated and measured return loss of the dual-band fleece fabric antenna.

both cases, and the measured return loss corresponds well with the simulations. The measured bandwidths also match well with the simulations. The measured lower resonant frequency was 1.927 GHz, and the higher resonant frequency was 2.450 GHz.

It is worth noting that the antenna fails to achieve the specified bandwidths of GSM1900 and WLAN bands. The required bandwidth for GSM1900 operation is from 1.850 to 1.990 GHz, and for WLAN operation, from 2.4 to 2.484 GHz. The measured bandwidth for the lower frequency band was from 1.876 to 1.967 GHz, and for the higher frequency band, from 2.414 to 2.491 GHz. The simulations show a similar behavior. The bandwidth was determined from $S_{11} = -6$ dB limit, which corresponds to a VSWR of less than 3. The bandwidths of the antenna remain too narrow because of fabric thickness. The simulations also predict that the dielectric should be thicker in order to obtain the required bandwidth.

The effect of fabric thickness on the impedance bandwidth was studied by constructing another antenna with a different fabric used as a dielectric. The second antenna was mounted on Vellux fabric. The relative permittivity of this fabric was close to that of the fleece fabric, but the thickness of the Vellux fabric was 5.5 mm. In order to compare the performance of the two antennas, the second antenna needed to be simulated and optimized for the correct dimensions. The dimensions of the second antenna were optimized to allow this antenna to operate in the 1.9 and 2.4 GHz bands. Figure 6.21 shows the measured return loss for both antennas.

The results show that thickness affects the impedance bandwidth as predicted. The second antenna with thicker fabric had wider bandwidth, nearly meeting the GSM1900 bandwidth requirements. The measured bandwidth for the lower fre-

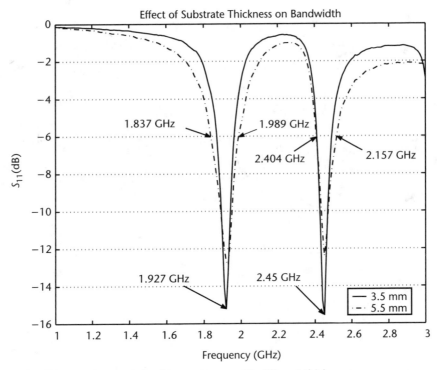

Figure 6.21 Measured return loss of two antennas with different thicknesses.

quency band was from 1.837 to 1.989 GHz, and for the higher frequency band, from 2.404 to 2.517 GHz. The results indicate the importance of proper fabric selection. Therefore, the desired requirements for a particular application need to be properly considered, and a suitable fabric needs to be selected to meet these requirements.

The radiation characteristics of the dual-band fleece fabric antenna were also studied. Figures 6.22 and 6.23 show the measured radiation patterns at the lower and higher resonant frequencies, respectively, for the E-plane and the H-plane (xz-plane and yz-plane). The peak antenna gain at the lower resonant frequency was approximately 9.26 dBi, and at the higher resonant frequency, it was 7.99 dBi. The half-power beamwidths in the E-plane were 64° at the lower resonant frequency, and 53° at the higher resonant frequency. In the H-plane, the corresponding values were 61° and 63°, respectively. The measured radiation patterns at both resonant frequencies corresponded well in both the E-plane and the H-plane. In the E-plane, the radiation patterns had very similar received power levels at both frequencies in the direction of the mainbeam. Both radiation patterns also had minima in similar directions.

6.7 Wearable Electromagnetic Bandgap Antenna: An Example of a WLAN Antenna

The size of a textile antenna can become an issue due to the low dielectric constant of a fabric material. However, the thickness of the antenna is not a major issue,

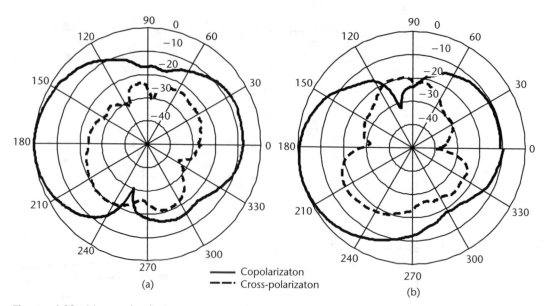

Figures 6.22 Measured radiation pattern at 1.92 GHz: (a) *xz*-plane, and (b) *yz*-plane. Solid line represents copolarization and dashed line represents cross-polarization.

since many outerwears can have thicknesses up to 10 mm, depending on the purpose of the clothing. In addition, most clothes are composed of different numbers of layered textiles. This provides new possibilities for antenna designers to develop multilayer antennas. Reviews of recent journal and conference publications show that useful applications of electromagnetic bandgap structures have emerged [38–40].

The term electromagnetic bandgap (EBG) has been used to describe various types of dielectric or metallic periodic structures. A complete EBG material is a periodic structure in which no electromagnetic propagation is possible within a certain frequency range for all angles and polarization states. An EBG structure forms a lattice whose period determines its resonant frequency—that is, the range of frequencies where the stopband exists for the transmission of microwave signals. This provides an effective method to suppress surface wave propagation and higher order harmonics in integrated microwave circuits and antennas. However, surface wave propagation is not an issue with textile antennas, but suppression of high-order harmonics can become a useful method.

It has been found that EBG structures can offer remarkable improvements over conventional microwave antennas and systems. For example, antenna radiation patterns, radiation efficiency, and impedance bandwidth can be improved with EBG structures, and even the antenna size can be reduced. However, EBG mushrooms have gained the most attention, in which an antenna is surrounded by metallic patches. There are many reports providing accurate information on how low-cost EBG mushrooms can be designed with patch antennas [41–44].

Figure 6.24 shows the geometry of a two-layer EBG textile antenna for WLAN. It has 6 × 6 EBG mushroom–like patches without vias on top of the first 4-mm-thick dielectric layer. The side length of each mushroom patch is 26 mm, and spacing between each is 2 mm. On top of the EBG mushroom is the next 4-mm-thick dielec-

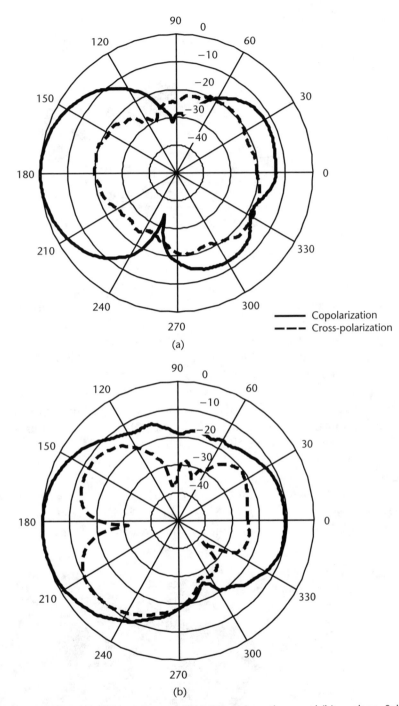

Figures 6.23   Measured radiation pattern at 2.45 GHz: (a) *xz*-plane, and (b) *yz*-plane. Solid line represents copolarization and dashed line represents cross-polarization.

tric layer, and the main radiating patch is on top of this layer. The dielectric material is conventional felt fabric with a dielectric constant of 1.1. A coaxial SMA connector was provided for the antenna feed. The feeding pin is not in contact with the EBG surface. The feed was located at a distance of 4 mm from the main patch edge

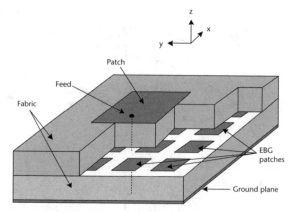

Figure 6.24   Geometry of the WEBGA. In this context EBG is referred to as the embedded patches in the substrate as shown in this figure.

in order to provide a good match. In actual applications, however, the SMA connector could be replaced by a more appropriate connector, such as microstrip line. Conductive parts (e.g., EBG mushroom, patch, and ground plane) are made out of copper tape. Figure 6.25 shows the photo of a measured prototype. In Figure 6.26, a photograph of an embedded patch surface is shown. Figure 6.27 shows the simulated S_{11} for a wearable WLAN antenna, with and without embedded patches. The result reveals that the resonant frequency is shifted down by 20% with the addition of embedded patches. The main patch size is 44 × 38 mm with EBG, and 52 × 46 mm without EBG. This means that the area of the EBG antenna is less than 70% of that of a conventional patch antenna because the EBG structure can be "hidden" inside the clothing. Another observation is that the embedded patch structure can be considered as an artificial textile material. In this example, the equivalent dielectric constant is 1.4, which is shown in Figure 6.27 as well. It is clear that in this contest the term *embedded patch* is more suitable than *electromagnetic bandgap* (EBG); however, in current literature these definitions are sometimes used interchangeably.

Figure 6.25   Photo of the measured prototype wearable electronic bandgap antenna (WEBGA).

Figure 6.26   Photo of the embedded patches.

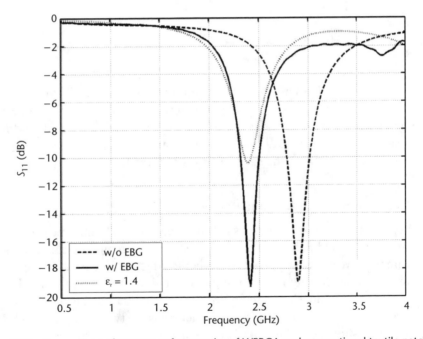

Figure 6.27   Comparison of resonance frequencies of WEBGA and conventional textile patch antenna (FDTD simulation results.)

The return loss was measured with an HP8510B network analyzer, and is shown in Figure 6.28. It can be seen that the center frequency is 2.55 GHz, slightly off the desired 2.45 GHz. However, the input-match bandwidth is 275 MHz, which is 100 MHz wider compared to conventional textile patch antenna. The input-match bandwidth is defined as VSWR 2:1, or when the S_{11} is not more than -10 dB. This shows that input-match bandwidth improvement is remarkable, with the added advantage a reduction of the antenna size.

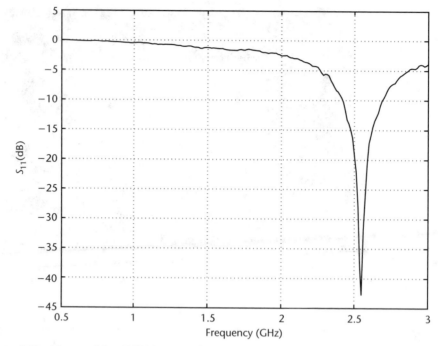

Figure 6.28 Measured S_{11} of WEBGA.

Radiation characteristics were simulated using the FDTD method developed at UCLA, and the results are shown in Figure 6.29 for the xz-plane and yz-plane. The directivity of the WEBGA is approximately 10 dB. Figures 6.30 and 6.31 show the measured radiation patterns in the xz-plane and yz-plane, respectively, and very good agreement with the simulations is observed.

6.7.1 Remarks on Antenna Bending

In general, in wearable systems flat antenna surfaces cannot be provided. Therefore, antennas should properly function even if the antenna is bent. Two textile WLAN antennas are again employed to study the effect of antenna bending, namely, a conventional patch antenna and the previously introduced wearable EBG antenna.

The test setup, shown in Figure 6.32, includes two plastic cylinders with diameters of 70 mm and 150 mm. These dimensions are typical for human body (e.g., arm, leg, and shoulder). Antennas are bent around the cylinder along two principal planes, xz and yz. The coordinates system is shown in Figure 6.24. Figures 6.33 and 6.34 show the results for the patch and EBG antennas, respectively for both bending diameters.

Table 6.3 summarizes the results of Figures 6.33 and 6.34 in terms of resonance frequency and input-match bandwidth deviation due to antenna bending. It can be observed that yz-plane bending has a minor effect on antenna performance compared to xz-plane bending. This is due to the fact that xz-plane bending affects on the antennas' resonance length. The more the antenna is bent (i.e., around a smaller diameter), the more resonance length is reduced, and thus it is shifted up. This is observable for both antennas. However, yz-plane bending effects the resonance frequency of the

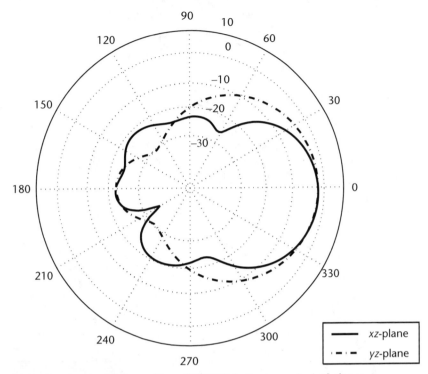

Figure 6.29   Simulated radiation pattern of WEBGA in the two principal planes.

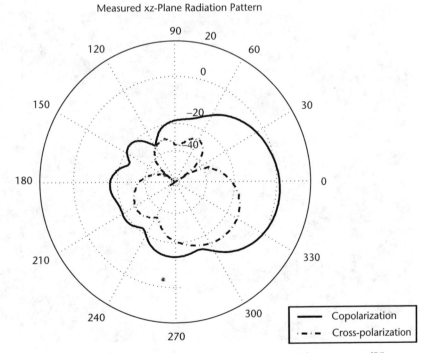

Figure 6.30   Measured *xz*-plane radiation pattern at 2.55 GHz (with respect to dBi).

Measured xz-Plane Radiation Pattern

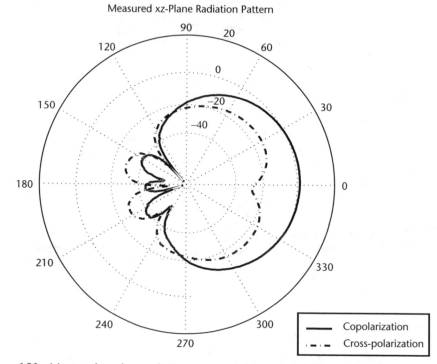

Figure 6.31   Measured *yz*-plane radiation pattern at 2.55 GHz (with respect to dBi).

Figure 6.32   Antenna-bending measurement setup at UCLA's anechoic chamber.

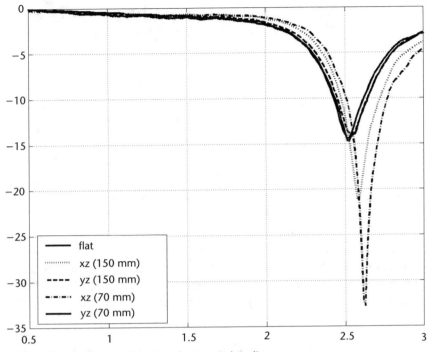

Figure 6.33   Measured S_{11} results of patch antenna bending.

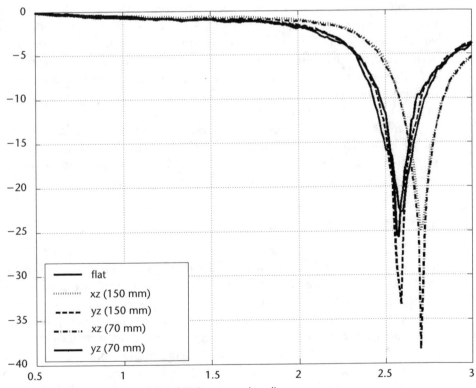

Figure 6.34   Measured S_{11} results of EBG antenna bending.

Table 6.3 Comparison of the Bending Effect of Patch and EBG Antennas

Bending Diameter	F_c (GHz)		Bandwidth (MHz)	
	Patch	EBG	Patch	EBG
Flat (0 mm)	2.51	2.58	212	287
xz (70 mm)	2.62	2.70	287	275
yz (70 mm)	2.51	2.59	237	275
xz (150 mm)	2.59	2.70	262	275
yz (150 mm)	2.55	2.59	212	287

conventional patch antenna as well. In terms of resonance frequency stability due to the bending, the EBG antenna performs better than the conventional patch antenna. Therefore, the EBG antenna easier to place within clothing, such as a sleeve, when an antenna's xz-plane is aligned with the sleeve length. Similar observations can be made for input-match bandwidth. The input-match bandwidth of patch antenna varies remarkably when the antenna is bent. However, the EBG antenna clearly outperforms the patch antenna, and the bandwidth deviation notably smaller.

6.8 Wearable Antennas Near the Human Body: An Example of a WLAN Antenna

Wearable antennas need to be designed to operate properly in the vicinity of the human body. In addition to this, special attention must be paid to the SAR, which aids in the quantitative study of power absorption issues. However, the use of cellular telephones and other personal communication services has increased research activities devoted to human-antenna interaction. Most of the research effort has been focused on the interaction between the handset antennas and the human head [45–49]. However, in the case of wearable antennas, the antennas are usually placed in the vicinity of the human torso or arm.

Here we address the following issues: (1) textile antenna input-match performance in the vicinity of the human body, (2) the radiation characteristics of the textile antenna, (3) the power absorbed in the human body, and (4) the peak and averaged SAR over 1g tissue in the 2.4-GHz ISM band.

6.8.1 Models and Methods

An FDTD algorithm developed at UCLA has been employed to model WLAN textile antennas for wearable applications. The textile antenna has a substrate material made out of conventional fleece fabric with a relative dielectric constant measuring from 1.1 to 2.4 GHz. The thickness of the fabric is 8 mm, providing adequate input-match bandwidth for the ISM band. The antenna has a ground plane and main radiating patch made out of knitted copper. The antenna performance was studied with a series of simulations, yielding an optimum patch size of 52 × 46 mm. The ground plane is 20 mm larger in each dimension. The feed is located 4 mm from the patch edge. In addition, the SAR performance of previously introduced textile EBG is discussed. Here the effect of vias on SAR performance is studied.

Two different human body models are used to study the textile antenna performance. The torso model is constructed from CT and MRI images of a real human body. Its cell size is 4 × 4 × 4 mm, containing a detailed organ model. This model is used to study the antenna performance when the antenna is located on the human chest. The other model is a simplified arm containing a 2-mm-thick skin layer, muscle, and bone. Anatomical features of the human body can be modeled by assigning a permittivity and conductivity to different tissues in the body and to each cell within the spatial grid. The electrical parameters corresponding to each tissue around the operating frequency 2.4 GHz have been obtained from [49], and their electrical parameters are listed in Table 6.4. The cell size in the arm is 1 × 1 × 1 mm. For comparison results, a half-wave dipole antenna is simulated in the same locations. In SAR computations, the average SAR is obtained by calculating the SAR over six neighboring cells. Antenna locations with respect to the human body model are shown in Figure 6.35.

6.8.2   Results

Figure 6.36 shows the theoretical results for the textile WLAN antenna, with and without the presence of the human torso or arm. The difference between these simulations is that the antenna was modeled with a coarser grid in the human torso, as explained in the previous section. Therefore, the feed location is not as accurate as in other cases. It can be seen that the textile WLAN performs very well in all cases. Therefore, the human body does not affect the input-match. Figures 6.37 and 6.38 show a similar study for xz-plane and yz-plane radiation patterns, respectively. As can be noted, there is a difference in the back radiation beamwidths. The directivity is close to 9 dB in all cases.

Having investigated the effect of the body on the electromagnetic characteristics of textile WLAN antennas, we now turn our focus to the rate of energy absorption, as defined by the SAR, which aids in the quantitative study of power absorption issues. Table 6.5 summarizes the absorbed power, maximum SAR, and 1-g averaged SAR for textile WLAN antennas. The results are compared with standard half-wavelength–long dipoles located at the same positions as textile WLAN antennas. The dipole distance from the human body was 4 mm. This result shows that the textile antenna performs significantly better. Figure 6.39 shows the normalized near field caused by the chest-attached antenna. The dark color outside the body indicates the strongest near-field values that are observed near the antenna. The figure also shows that the near field is weak inside the body.

Finally, we turn our focus on how EBG mushroom vias affect the SAR performance of the previously introduced textile EBG antenna. The same human body model has been employed together with the EBG antenna attached onto the chest.

Table 6.4   Arm Model Parameters at 2.4 GHz

Tissue	ε_r	σ (S/m)	$\rho(*10^3$ kg/m$^3)$
Skin	38.0629	1.4408	1.01
Muscle	53.6391	1.77472	1.04
Bone	15.0087	0.586055	1.85

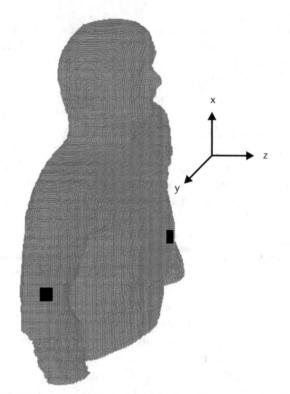

Figure 6.35 Human torso for wearable antenna modeling. The antenna locations are shown in black.

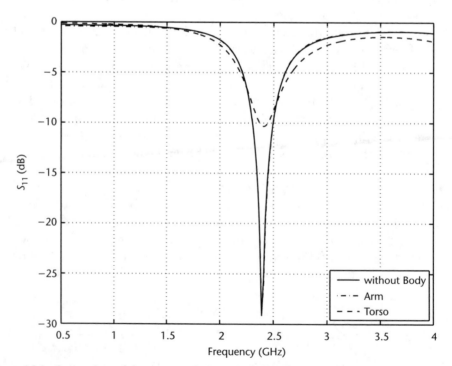

Figure 6.36 Comparison of the resonant frequency of a textile antenna, with and without the presence of a human body.

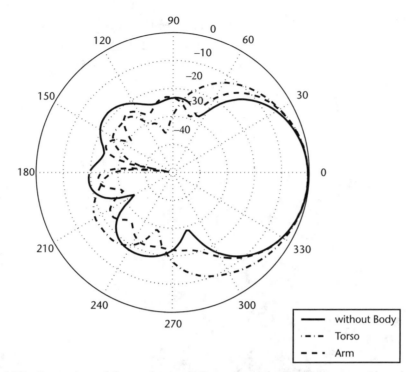

Figure 6.37   Comparison of the *xz*-plane radiation pattern of a textile antenna, with and without the presence of a human body.

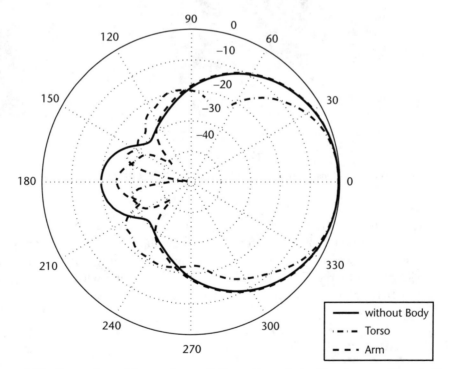

Figure 6.38   Comparison of the *yz*-plane radiation pattern of a textile antenna, with and without the presence of a human body.

Table 6.5   Computed SAR Values at 2.4 GHz for 1-W Delivered Power

Antenna	P_{abs} (mW)	SAR_{peak} (mW/g)	SAR_{avg} (mW/g)
Patch + Torso	12	0.054	0.00054
Dipole + Torso	124	0.655	0.0056
Patch + Arm	21	0.25	0.0043
Dipole + Arm	750	104.5	0.15

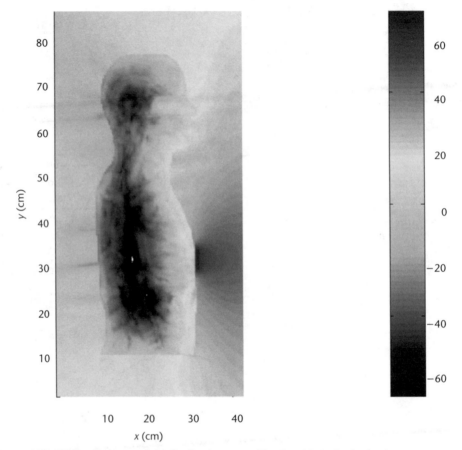

Figure 6.39   Normalized near-field distribution caused by the chest-attached antenna.

Table 6.6 summarizes the SAR performance for the EBG antenna with and without vias at 2.45 GHz. The computed results show that the addition of vias decreases the absorbed power, and thus reduces SAR. Figure 6.40 shows that the normalized near field caused by the chest attached textile EBG antenna without vias. The strongest near-field values are observed near the antenna.

6.9   Conclusions

With a variety of promising approaches and the involvement of leading research groups, wearable textile antennas will emerge and replace conventional rigid anten-

Table 6.6 Comparison of the Effect of Vias on EBG Antenna SAR Performance

	Without Vias	*With Vias*
P_{abs} (mW)	17	13
SAR_{peak} (mW/g)	0.34	0.12
SAR_{avg} (mW/g)	0.00019	0.00014

Note: SAR values are computed at 2.45 GHz for 1-W delivered power.

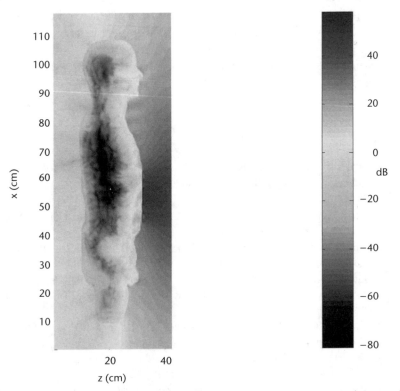

Figure 6.40 Electric near-field distribution due to EBG antennas without vias is shown at 2.45 GHz (normalized at 1-W delivered power).

nas in different applications. Current research on textile antennas and the knowledge of conventional microstrip antennas will provide a solid foundation for the improved antenna designs based on the technologies highlighted here. In addition to the further development of existing technologies, new solutions are likely to emerge.

One such new technology could be the so-called system-on-fabric (SoF), in which conventional printed circuit boards (PCB) are replaced by textile-like materials, allowing further system-level integration. Within SoF, integrated circuits (IC) and multichip modules (MCM) can be directly attached onto textile or even onto biomaterials. This creates new methods for designing intelligent textiles or spare parts for vital organs in the human body.

There is still an enormous amount of research and development that needs to be performed in this area in order to reach an optimal performance level. The multidisciplinary field of wearable antennas combines electromagnetics, materials sciences, electronics circuit design and miniaturization, and bioelectronics. The self-organizing materials may give new insight into wearable antenna development. This area will continue to benefit from progress in numerical methods and computing power, as well as from any decrease in computing cost.

Acknowledgments

The authors would like to thank Dr. Fan Yang, Mr. Jaehoon Kim, Mr. Majid Manteghi from the UCLA Antenna Laboratory, Dr. Mikko Keskilammi, Ms. Heli Hurme, and Ms. Marijke Schaffrath from Tampere University of Technology, Institute of Electronics, for fruitful discussions and effective collaborations. In addition, we would like to thank Reima Clothing for allowing us to use their smart clothing picture.

References

[1] Marculescu, D., et al., "Ready to Ware," *IEEE Spectrum*, October 2003, pp. 28–32.

[2] Lebaric, J., and A.-T. Tan, "Ultra-Wideband RF Helmet Antenna," *IEEE MILCOM 2000*, Vol. 2, 2000, pp. 591–594.

[3] Lebaric, J., and A.-T. Tan, "Ultra-Wideband Conformal Helmet Antenna," *IEEE Asia Pacific Microwave Conference*, 2000, pp. 1477–1481.

[4] Abramo, R., et al., "Fabrication and Testing of the COMVIN Vest Antenna," *IEEE MILCOM 2000*, Vol. 2, 2000, pp. 595–598.

[5] Lebaric, J. E., R. W. Adler, and M. E. Limbert, "Ultra-Wideband, Zero Visual Signature RF Vest Antenna for Man-Portable Radios," *IEEE MILCOM 2001*, Vol. 2, 2001, pp. 1291–1294.

[6] Kohls, E. C., et al., "A Multi-Band Body-Worn Antenna Vest," *2004 IEEE Antennas and Propagation Society Intl. Symp.*, Vol. 1, Monterey, CA, 2004, pp. 447–450.

[7] Fear, E. C., et al., "Enhancing Breast Tumor Detection with Near-Field Imaging," *IEEE Microwave Magazine*, March 2002, pp. 49–56.

[8] Hagness, S. C., A. Taflove, and J. E. Bridges, "Three-Dimensional, FDTD Analysis of a Pulsed Microwave Confocal System for Breast Cancer Detection: Design of an Antenna-Array Element," *IEEE Trans. on Antennas and Propagation*, Vol. 47, No. 5, 1999, pp. 783–791.

[9] Hagness, S. C., A. Taflove, and J. E. Bridges, "Three-Dimensional FDTD Analysis of an Ultrawideband Antenna-Array Element for Confocal Microwave Imaging of Nonpalpable Breast Tumors," *IEEE Int. Symp. Antennas and Propagation*, Vol. 3, 1999, pp. 1886–1889.

[10] Lund, D. V., "Investigation of the Behaviour of Microwave Thermography Antennas When Coupled to Layered Tissue Regions," *IEE Colloquium on Applications of Microwaves in Medicine*, 1995, pp. 7/1–7/5.

[11] Smailagic, A., and D. P. Siewiorek, "A Case Study in Embedded-System Design: The VuMan 2 Wearable Computer," *IEEE Design and Test Computers*, Vol. 10, No. 3, September 1993, pp. 56–67.

[12] Rantanen, J., et al., "Smart Clothing for the Arctic Environment," *IEEE 4th Int. Symp. Wearable Computers*, 2000, pp. 15–23.

[13] Kukkonen, K., et al., "The Design and Implementation of Electrically Heated Clothing," *IEEE 5th Int. Symp. Wearable Computers*, 2001, pp. 180–181.

[14] Rantanen, J., et al., "Improving Thermal Comfort with Smart Clothing," *IEEE Int. Conference on Systems, Man, and Cybernetics*, Vol. 2, 2001, pp. 795–800.

[15] http://www.reimasmart.com.

[16] StarTiger2 home page http://www.esa.int/export/esaCP/SEM3GOGHZTD_Improving_0.html, cited February 7, 2005.

[17] Salonen, P., et al., "A Small Planar Inverted-F Antenna for Wearable Applications," *Third Int. Symp. Wearable Computers*, 1999, pp. 95–100.

[18] Salonen, P., et al., "A Novel Antenna System for Man-Machine Interface," *ICMA 2000, Int. Conference on Machine Automation*, Osaka, Japan, September 2000.

[19] Salonen, P., and J. Rantanen, "A Dual-Band and Wide-Band Antenna on Flexible Substrate for Smart Clothing," *27th Annual Conference of the IEEE Industrial Electronics Society (IECON 2001)*, Denver, CO, 2001.

[20] Salonen, P., M. Keskilammi, and L. Sydänheimo, "Antenna Design for Wearable Applications," *World Multiconference on Systemics, Cybernetics and Informatics*, Vol. 4, 2001, pp. 496–502.

[21] Salonen, P., M. Keskilammi, and L. Sydänheimo, "A Low-Cost 2.45 GHz Photonic Band-Gap Patch Antenna for Wearable Systems," *IEE 11th Int. Conference on Antennas and Propagation (ICAP 2001)*, Vol. 2, No. 480, Manchester, U. K., 2001, pp. 719–723.

[22] Massey, P. J., "Mobile Phone Antennas Integrated Within Clothing," *IEE 11th Int. Conference on Antennas and Propagation (ICAP 2001)*, Vol. 1, No. 480, Manchester, U. K., 2001, pp. 344–347.

[23] Salonen, P., and L. Sydänheimo, "Development of an S-Band Flexible Antenna for Smart Clothing," *2002 IEEE Antennas and Propagation Society Int. Symp.*, San Antonio, TX, 2002.

[24] Hurme, H., P. Salonen, and J. Rantanen, "On the Study of Antenna Placement in a Smart Clothing," in *Proc. IASTED Int. Conference; Modeling and Simulation*, M. H. Hamza, (ed.), Palm Springs, CA, February 24–26, 2003, pp. 1–6.

[25] Salonen, P., and H. Hurme, "Modeling of a Fabric GPS Antenna for Smart Clothing," in *Proc. IASTED Int. Conference; Modeling and Simulation*, M. H. Hamza, (ed.), Palm Springs, CA, February 24–26, 2003, pp. 18–23.

[26] Salonen, P., and H. Hurme, "A Novel Fabric WLAN Antenna for Wearable Applications," *2003 IEEE Antennas and Propagation Society Int. Symp.*, Columbus, OH, Vol. 2, 2003, pp. 700–703.

[27] Salonen, P., et al., "Effect of Conductive Material on Wearable Antenna Performance: A Case Study of WLAN Antennas," *2004 IEEE Antennas and Propagation Society International Symp.*, Vol. 1, Monterey, CA, 2004, pp. 455–458.

[28] Salonen, P., et al., "Effect of Textile Materials on Wearable Antenna Performance: A Case Study of GPS Antennas," *2004 IEEE Antennas and Propagation Society International Symp.*, Vol. 1, Monterey, CA, 2004, pp. 459–462.

[29] Salonen, P., et al., "Dual-Band Wearable Textile Antenna," *2004 IEEE Antennas and Propagation Society International Symp.*, Vol. 1, Monterey, CA, 2004, pp. 463–466.

[30] Salonen, P., et al., "WEBGA—Wearable Electromagnetic Band-Gap Antenna," *2004 IEEE Antennas and Propagation Society International Symp.*, Vol. 1, Monterey, CA, 2004, pp. 451–454.

[31] Salonen, P., Y. Rahmat-Samii, and M. Kivikoski, "Wearable Antennas in the Vicinity of Human Body," *2004 IEEE Antennas and Propagation Society International Symp.*, Vol. 1, Monterey, CA, 2004, pp. 467–470.

[32] Tanaka, M., and J.-H. Jang, "Wearable Microstrip Antenna," *2003 IEEE Antennas and Propagation Society Int. Symp.*, Columbus, OH, Vol. 2, 2003, pp. 704–707.

[33] Wong, T. C. P., et al., "Fabrication and Evaluation of Conducting Polymer Composite as Radar Absorbers," *IEE 8th Int. Conference on Antennas and Propagation (ICAP 1993)*, Vol. 2, 1993, pp. 934–937.

[34] Wong, T. C. P., et al., "Characterization of Conducting Polymer-Loaded Composite Materials at Oblique Incidence and Their Application in Radar Absorbers," *IEE 9th Int. Conference on Antennas and Propagation (ICAP 1995)*, Vol. 2, No. 407, 1995, pp. 441–444.

[35] Callister, Jr., W. D., *Materials Science and Engineering; An Introduction*, 6th ed., New York: John Wiley & Sons, 2003.

[36] Delinova 200 of 1110 dtex DuPont's Cordura, Data sheet, Delcotex, Germany, 2003.

[37] Ingalls, M., and D. Smith, "Microstrip Antennas for GPS Applications," *IEEE Position Location and Navigation Symp.*, 2002, pp. 20–27.

[38] Special Issue on Metamaterials, *IEEE Trans. on Antennas and Propagation*, Vol. 51, No. 10, October 2003.

[39] *IEEE Trans. on Microwave Theory and Techniques*, Vol. 47, No. 11, November 1999.

[40] *Proc. 24th ESTEC Antenna Workshop Innovative Periodic Antennas: Photonic Bandgap, Fractal and Frequency Selective Structures*, Noordwijk, the Netherlands, May 30–June 1, 2001.

[41] Gonzalo, R., P. de Maagt, and M. Sorolla, "Enhanced Patch-Antenna Performance by Suppressing Surface Waves Using Photonic-Bandgap Substrates," *IEEE Trans. on Microwave Theory and Techniques*, Vol. 47, No. 11, November 1999, pp. 2131–2138.

[42] Yang, F., and Y. Rahmat-Samii, "Reflection Phase Characterization of an Electromagnetic Band-Gap (EBG) Surface," *IEEE Antennas and Propagation Symp.*, Vol. 3, 2002, pp. 744–747.

[43] Yang, F., and Y. Rahmat-Samii, "Reflection Phase Characterizations of the EBG Ground Plane for Low Profile Wire Antenna Applications," *IEEE Trans. on Antennas and Propagation*, Vol. 51, No. 10, October 2003, pp. 2691–2703.

[44] de Maagt, P., et al., "Electromagnetic Bandgap Components and (Sub)Millimeter Wave Applications," *IEEE Trans. on Antennas and Propagation*, Vol. 51, No. 10, October 2003, pp. 2667–2677.

[45] Jensen, M. A., and Y. Rahmat-Samii, "EM Interaction of Handset Antennas and a Human in Personal Communications," *IEEE Proc.*, Vol. 83, 1995, pp. 7–17.

[46] Iskander, M. E., Y. Zhengqing, and R. Quintero-Illera, "Polarization and Human Body Effects on the Microwave Absorption in a Human Head Exposed to Radiation from Handheld Devices," *IEEE Trans. on Microwave Theory and Techniques*, Vol. 48, No. 10, November 2000, pp. 1979–1987.

[47] Dimbylow, P. J., and O. P. Gandhi, "Finite-Difference Time-Domain Calculations of SAR in a Realistic Heterogeneous Model of the Head for Plane Wave Exposure from 600 MHz to 3 GHz," *Physics in Medicine and Biology*, Vol. 36, August 1991, pp. 1075–1089.

[48] Rahmat-Samii, Y., et al., "Antennas and Humans in Personal Communications," in *Mobile Antenna Systems Handbook*, K. Fujimoto and J. R. James, (eds.), 2nd ed., Chap. 7, Norwood, MA: Artech House, 2000.

[49] http://www.fcc.gov/fcc-bin/dielec.sh.

Body-Sensor Networks for Space and Military Applications

Rainee N. Simons

7.1 Introduction

The robotic and human space exploration program at NASA seeks to develop implantable wireless sensors to monitor the physiological parameters of humans during space flights [1]. This focus is rather unique when compared to efforts by other investigators, which have been mainly in the area of electromagnetic wave interaction with biological systems, hyperthermia treatment of cancer, and RF and microwave applications in medical treatment and biological effects [2–6]. Conventional sensors that are used in biomedical implants require powering through batteries and lead wires. The disadvantages of this approach are the restriction of mobility, the requirement for shielding from moisture, the potential for malfunctioning, the possible risk for infection, and the limited lifespan of the sensor.

This chapter presents an overview of a typical biomedical RF telemetry system. In addition, the design of an antenna for contactless powering and RF telemetry is discussed. The contactless powering and RF telemetry are for powering and data acquisition from the implantable biomicroelectromechanical systems (bio-MEMS) sensor. Furthermore, applications of these sensors in military and space are presented as examples. The approach relies on active inductive coupling or magnetic induction between the implanted sensor and the printed antenna in the interrogating/receiving handheld device. RF telemetry reception from the implanted sensors has been demonstrated in [7–11], and Table 7.1 summarizes the dimensions of the implanted on-chip inductor/antenna, as well as the distance over which they communicated. However, the approach presented in [11] and in this chapter has the following unique features. First, the size of our inductor/antenna is significantly smaller (1 × 1 mm), resulting in smaller implant size. Second, an MMIC amplifier can be integrated with the pickup antenna in the handheld device to enable communications across larger implant depths. When compared with conventional sensors, the sensors with telemetry have the following advantages. First, the size of the inductor/antenna is very small, which allows the device to be integrated into miniature bioimplants, such as MEMS pressure sensors. Second, these sensors eliminate the need to implant batteries, which reduces the possibility of infection. Third, the circuit operates only when interrogated by an external handheld device, which min-

189

imizes power dissipation in the biological tissue, avoids local heating, and extends the lifespan of the sensor. Fourth, these sensors eliminate feedthrough wires for powering and telemetry, which greatly enhances mobility and reduces the risk of infection. The frequency band that is available on an unlicensed basis for body implants extends from 402 to 405 MHz, and has a maximum allowable power level of 25 μW external to the body. This combination of frequency and power allows an operating range of approximately 2m.

Prior publications by the author [11–16] have reported on the wireless RF telemetry scheme, the notional implantable bio-MEMS sensor, the validation of the telemetry concepts using biological tissue-like phantom media, and the radiation characteristics of the antenna in the implantable sensor and in the handheld unit. In the sections that follow, a typical biosensor system and the basics of biomedical telemetry are introduced, assuming that the implant is a bio-MEMS–based capacitive pressure sensor. In addition, it is assumed that the implant uses a microinductor/antenna for contactless powering and RF telemetry. This RF telemetry concept can be extended to the other types of sensor systems presented in Table 7.1.

7.2 Biosensor System and Basics of Biomedical RF Telemetry

A typical biosensor system consists of a miniature implanted sensor with an integrated antenna, and a larger external antenna integrated into a handheld device. The implanted antenna and the external antenna transmit and receive the telemetry signal from the sensor. The implanted antenna also enables inductive or contactless powering of the biosensor.

The contactless powering and telemetry concept, including the miniature square spiral inductor/antenna circuit intended for integration with a MEMS pressure sensor, is illustrated in Figure 7.1(a). The pressure sensor is of the capacitive type, and is located in the annular region of the inductor. The inductor behaves both as an inductance as well as an antenna, thereby allowing the sensor to receive as well as

Table 7.1 Summary of Implantable Inductor/Antenna Dimensions, Operating Frequency, Link Distance, Implant Power, and Sensor Application

On-Chip Inductor Dimensions (mm)	Wireless Link Operating Frequency (MHz) and Distance (mm)	Implant Power Consumption	Sensor Application	Reference
2 × 10	4, 30	3 mA at 4V	Nerve stimulation system	Von Arx and Najafi [7]
5 × 5	4, 5	100 μA at 3.5V	Measuring intracranial pressure	Eggers et al. [8]
10.3 diameter	13.56, 30	280 μW and 3.5V 240 μW	1. Measuring intraocular pressure 2. Measuring blood pressure intravascular 3. Stimulating nerve cells of retina	Ullerich et al. [9] Mokwa and Schnakenberg [10]
1 × 1	≈ 300, 100	N/A	Measuring pressure	Simons et al. [11]

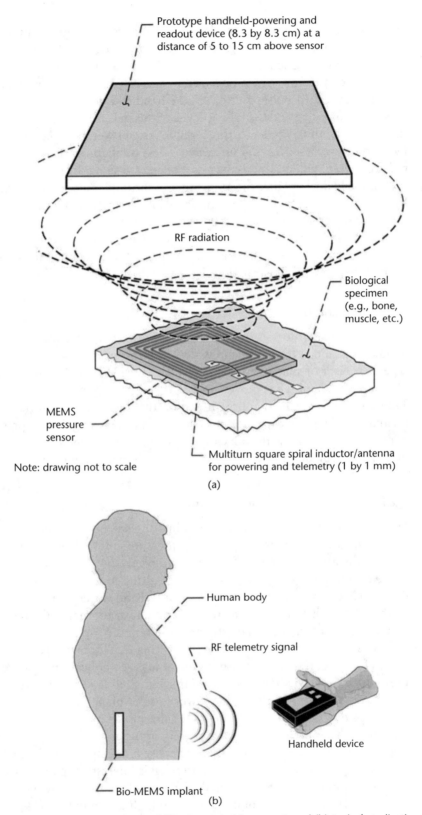

Figure 7.1   Contactless powering and RF telemetry: (a) concept, and (b) typical application in biosensors.

radiate energy. In the receive mode, the inductance picks up energy and charges the MEMS pressure sensor diaphragm capacitance. In the transmit mode, the above inductance and capacitance form a parallel resonant circuit and radiate energy through the inductor, which now behaves as a planar spiral antenna. To obtain a pressure reading, a pulse emitted by the handheld device initially interrogates the sensor. At the rising and the falling edges of this pulse, a voltage is induced in the spiral inductor, thus implementing contactless powering. The waveform of this induced voltage is a decaying sine wave. These oscillations also cause the inductor to radiate energy, which is picked up as a telemetry signal by the receiving antenna in the handheld device. Since the inductance of the implanted sensor circuit is fixed, the frequency of the decaying sine wave is mainly determined by the capacitance introduced by the pressure sensor. Thus, the larger the pressure difference, the larger the frequency offset between the received telemetry in the two pressure states. The implanted bio-MEMS sensor and the handheld device together form the wireless RF telemetry system, as illustrated in Figure 7.1(b).

7.2.1 Implantable Pressure Sensor

A typical implantable pressure sensor consists of a diaphragm suspended over a cavity micromachined from a silicon wafer, and the sensor is typically of the capacitive type. A trilayer of silicon dioxide and silicon nitride is used to manufacture the diaphragm. The diaphragm moves up and down in response to mechanical pressure. Thin gold films on the diaphragm and on the lower surface of the cavity together form a parallel plate capacitor, whose capacitance changes with pressure. For this application, a sensor with capacitance change in the range from 0.3 to 4 pF is adequate. A schematic of this pressure sensor is shown in Figure 7.2.

7.2.2 Integrated Inductor/Antenna

Figure 7.3(a) shows a schematic of a miniature square spiral inductor/antenna that can be integrated into the periphery of a pressure sensor. The strip and separation or gap widths are indicated as W and G, respectively. The outer dimensions of the inductor are approximately 1×1 mm, and the inductor is fabricated on a good dielectric, such as a high-resistivity silicon (HR-Si) wafer, to reduce the attenuation of the signals. The above dimensions and substrate materials are typical for an implantable sensor. An initial estimate based on the capacitance values of the pressure sensor show that an inductance (L) with a quality factor (Q) of approximately 150 nH and 10, respectively, are adequate for the application. The frequency range over which this device can operate is approximately 200 to 700 MHz. In a practical circuit, the presence of a parasitic lower ground plane inadvertently introduced by the capacitive pressure sensor degrades the inductance and quality factor of the inductor. This is because the image current in the ground planes flows in a direction opposite to the current on the spiral, thereby reducing the magnetic field and thus the overall inductance [17, 18]. In contrast, it is interesting to observe that the inductance of the inductor with a serrated ring ground plane, as shown in Figure 7.3(b), is about the same as that of an inductor without a ground plane [14]. This is because the slots in the serrated ground plane act as open circuits, suppressing the flow of image currents [17, 18].

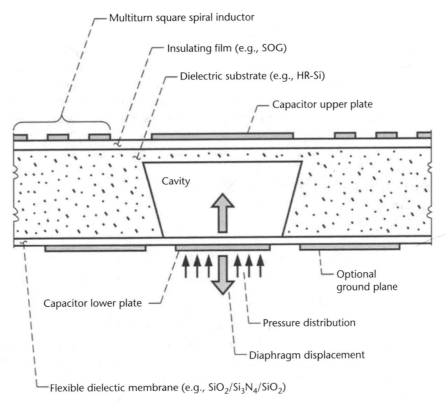

Multiturn square spiral inductor

Insulating film (e.g., SOG)

Dielectric substrate (e.g., HR-Si)

Capacitor upper plate

Cavity

Optional ground plane

Capacitor lower plate

Pressure distribution

Diaphragm displacement

Flexible dielectic membrane (e.g., $SiO_2/Si_3N_4/SiO_2$)

Figure 7.2   Cross-sectional view of the capacitive pressure sensor.

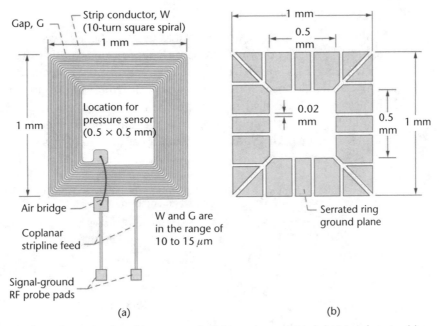

Gap, G

Strip conductor, W (10-turn square spiral)

1 mm

Location for pressure sensor (0.5 × 0.5 mm)

1 mm

Air bridge

Coplanar stripline feed

W and G are in the range of 10 to 15 μm

Signal-ground RF probe pads

(a)

1 mm

0.5 mm

0.02 mm

0.5 mm

1 mm

Serrated ring ground plane

(b)

Figure 7.3   Miniature implantable square spiral chip antenna on a dielectric substrate: (a) square spiral conductors, and (b) optional serrated ring ground plane.

7.2.3 External Pickup Antenna

The pickup antenna in the handheld device is a printed multiturn loop antenna, as illustrated in Figure 7.4. For high sensitivity, the input impedance of the loop antenna is matched to the input impedance of a MMIC low-noise amplifier (LNA) chip in the receiver. The complete handheld device, which houses the pickup antenna and other signal processing circuitry, is schematically illustrated in Figure 7.5.

7.3 Antenna Design for Body Sensors

In this section, general design guidelines for antennas integrated with body sensors are presented. These include the radiation resistance, loss resistance, and the inductance of the antenna. An equivalent circuit model is also discussed.

7.3.1 Implantable Antennas

Several researchers have demonstrated RF antennas for inductive powering and data communications in implantable biosensors [9, 15, 19–22]. These implantable biosensor antennas are manufactured in various shapes and sizes. Their design, construction, operating frequency, and intended biomedical applications are also different. The key features of these antennas are presented in Table 7.2. In spite of their differences, these antennas fall into one or two main categories, and their designs are presented in the next two sections.

7.3.1.1 Miniature Loop

Several antennas in Table 7.2 are fabricated by winding a very small diameter wire into a circular coil. The overall conductor length in this case is usually small compared to the operating wavelength. In such situations, the antenna can be modeled as a small circular loop with uniform current distribution, as illustrated in Figure 7.6. The uniform current distribution approximation along the circumference of the loop is valid only when the circumference is less than approximately 0.2λ, where λ is the free-space wavelength. Numerical examples at the end of this section present the

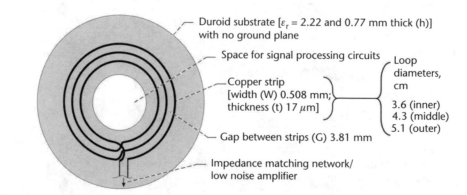

Figure 7.4 Printed multiturn loop antenna on a dielectric ring substrate.

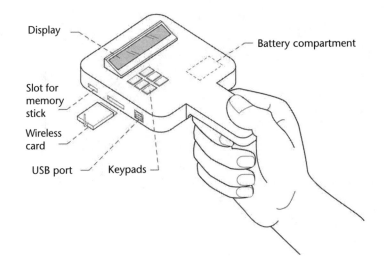

Figure 7.5   Schematic of the complete handheld device, which houses the multiturn printed loop antenna, as well as the signal processing circuits.

Table 7.2   Summary of State-of-the-Art Antenna Features in Implantable and Inductively Powered Biosensors

Type of Antenna and Dimensions (mm)	Operating Frequency (MHz) and Function	Biosensor Application	Reference
Off-the-shelf chip inductor,* 1.4 × 2.2 × 1.45	40; Data transmission	Ingestible devices	Ahmadian et al. [19]
Planar circular microcoil on polyimide foil, diameter 10.3	13.56; Data transmission	Intraocular pressure sensor	Ullerich et al. [9]
Manually wound circular coil, diameter 6.0	Unknown; Data reception	Visual prosthesis for epiretinal stimulation	Mokwa [20]
Rectangular patch on Teflon glass, 12.5 × 8.5	220; Data transmission	Strain monitoring in orthopedic implants	Van Schuylenbergh [21]
Wound circular coil, diameter 4.7	1; Data transmission	Recording peripheral neural signals from axons	Akin [22]
Planar square spiral on high-resistivity silicon, 1 × 1	235; Data transmission	Spinal implant	Simons [15]

*Battery powered.

results of a trade-off study between the loop radius and the frequency of operation as parameters on the inductive power in the near field and the radiation resistance of the antenna.

The magnetic field and electric field components associated with the small loop in Figure 7.6 are written as [23]

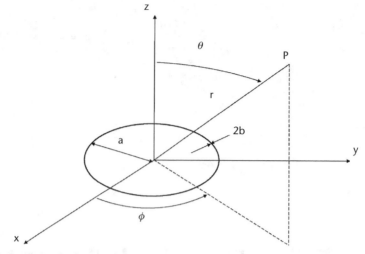

Figure 7.6 Small circular loop with uniform current distribution.

$$H_r = j\frac{ka^2 I_0 \cos\theta}{2r^2}\left[1+\frac{1}{jkr}\right]e^{-jkr} \quad A/m \tag{7.1}$$

$$H_\theta = -\frac{(ka)^2 I_0 \sin\theta}{4r}\left[1+\frac{1}{jkr}-\frac{1}{(kr)^2}\right]e^{-jkr} \quad A/m \tag{7.2}$$

$$H_\phi = 0 \tag{7.3}$$

$$E_r = E_\theta = 0 \tag{7.4}$$

$$E_\phi = \eta\frac{(ka)^2 I_0 \sin\theta}{4r}\left[1+\frac{1}{jkr}\right]e^{-jkr} \quad V/m \tag{7.5}$$

where $k = 2\pi/\lambda$, a is the loop radius in meters, and I_0 is current magnitude in amperes.

The complex power in watts along a radial direction away from a small loop antenna is given by [23]

$$P_r = \eta\left(\frac{\pi}{12}\right)(ka)^4|I_0|^2\left[1+j\frac{1}{(kr)^3}\right] \tag{7.6}$$

where η is the intrinsic impedance, equal to 376.73, or approximately $120\pi\Omega$. Observe that for small values of kr ($kr \ll 1$), the second term within the square brackets of (7.6) is dominant, which makes the power mainly inductive. This reasoning leads us to believe that the radial magnetic energy is larger than the electric energy, which is successfully exploited in the design of biotelemetry systems.

Example 7.1   As a numerical example, consider an implantable biosensor with a loop antenna having a radius of 2.0 mm and operating at a frequency of 50 MHz. Assume that I_0 is equal to 1A, and the distance r is equal to 10 cm. From (7.6), $P_r =$ $1.9 \times 10^{-9} + j1.65 \times 10^{-6}$ W. Notice that the reactive power is greater than the real power by three orders of magnitude. Thus, magnetic coupling is invariably used for RF telemetry in biosensor systems.

The radiation resistance in ohms of the small loop antenna is given by [23]

$$R_r = 20\pi^2 \left(\frac{C}{\lambda}\right)^4 \tag{7.7}$$

where C is the circumference of the loop, equal to $2\pi a$.

Example 7.2   For the loop in Example 7.1, the radiation resistance from (7.7) is $3.8 \times 10^{-9}\ \Omega$.

If the loop has N turns, then the radiation resistance given by (7.7) is multiplied by N^2. This simple feature of a loop antenna is very attractive, and is quite often exploited in the design of biotelemetry systems.

Example 7.3   In the above examples, if the radius of the loop is doubled to 4.0 mm, and the frequency remains the same, then the inductive part of P_r and the radiation resistance are $j2.65 \times 10^{-5}$ W and $6.1 \times 10^{-8}\ \Omega$, respectively. With respect to the original loop, it is observed that the near-field inductive power, as well as the radiation resistance, is higher by a factor of 16.

As a third case, if the radius of the loop remains as 2.0 mm while the operating frequency is doubled to 100 MHz, then the inductive part of P_r and the radiation resistance are $j3.31 \times 10^{-6}$ W and $6.1 \times 10^{-8}\ \Omega$, respectively. Comparing this set of values with the original values, it is observed that the near-field inductive power is higher only by a factor 2, while the radiation resistance is higher by a factor of 16.

As a last case, if both the radius of the loop as well as the operating frequency were doubled to 4.0 mm and 100 MHz, respectively, then the inductive part of P_r and the radiation resistance are $j5.3 \times 10^{-5}$ W and $9.7 \times 10^{-7}\ \Omega$, respectively. Again comparing with the original loop, the near-field inductive power is higher by a factor 32, and the radiation resistance is higher by a factor of 256.

Most often, we do not have the freedom to arbitrarily change the frequency of operation, which is decided by regulatory considerations. However, we are able to choose the dimensions of the antenna. The above examples demonstrate that a small increase in the radius of the implanted loop can significantly enhance the near-field inductive power and thus improve the signal-to-noise ratio, permitting communications at a greater distance or through thicker tissue layers.

The miniature loop antenna is represented for the purpose of computer modeling by a lumped equivalent circuit model, as shown in Figure 7.7. The input impedance Z_{in} of the loop is given by [23]

$$Z_{in} = R_{in} + jX_{in} = (R_r + R_L) + j(X_A + X_i)\quad \Omega \tag{7.8}$$

where:

R_r is the radiation resistance given by (7.7) in ohms

R_L is the loss resistance of the loop conductor $= \left(\dfrac{\text{length}}{\text{perimeter}}\right)R_s = \left(\dfrac{a}{b}\right)R_s$

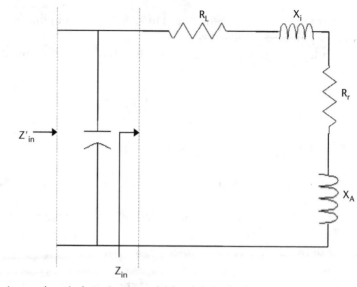

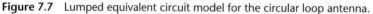

Figure 7.7 Lumped equivalent circuit model for the circular loop antenna.

b is the loop wire radius in meters

R_s is the surface resistance = $\sqrt{\dfrac{\pi f \mu_0}{\sigma}}$ Ω/Square

f is the frequency of operation in hertz

$\mu_0 = 4\pi \times 10^{-7}$ H/m

σ is the conductivity of the loop conductor (for copper, $\sigma = 5.7 \times 10^7$ S/m)

X_A is the inductive reactance of the loop antenna, equal to $2\pi f L_A$

X_i is the reactance of loop conductor, equal to $2\pi f L_i$

The inductive reactance X_A of the loop is calculated from the inductance L_A. The inductance L_A is also referred to as external inductance [24]. For a circular loop of radius a and wire radius b, L_A is given by [23]

$$L_A = \mu_0 a \left[\ln\left(\frac{8a}{b}\right) - 2 \right] \quad H \tag{7.9}$$

For a square loop with sides a and wire radius b, L_A is given by [23]

$$L_A = 2\mu_0 \frac{a}{\pi} \left[\ln\left(\frac{a}{b}\right) - 0.774 \right] \quad H \tag{7.10}$$

The inductive reactance X_i of the loop conductor is calculated from the inductance L_i, and is given by [23]

$$L_i = \frac{1}{2}\mu_0 \pi a \quad H \tag{7.11}$$

The inductance L_i is also referred to as the internal inductance [24].

In Figure 7.7, the capacitor C_s, which could be a combination of the distributed stray capacitance as well as a tuning capacitance, is used to resonate the antenna, and is chosen as [23]

$$C = \frac{1}{2\pi f} \frac{X_{in}}{R_{in}^2 + X_{in}^2} \quad F \tag{7.12}$$

At resonance, the input impedance Z'_{in} is given by [23]

$$Z'_{in} = R_{in} + \frac{X_{in}^2}{R_{in}} \quad \Omega \tag{7.13}$$

Example 7.4   As a numerical example, consider an implantable biosensor with a single turn loop antenna having a radius of 4.0 mm, constructed from a very thin wire of radius 0.1 mm, and operating at a frequency of 50 MHz. Then, from (7.7) and (7.8), R_r and R_L are equal to 6.077×10^{-8} Ω and 7.444×10^{-2} Ω, respectively. In addition, from (7.9) and (7.11), L_A and L_i are 1.894×10^{-8} H and 7.896×10^{-9} H, respectively. Substituting these values into (7.8), Z_{in} equals $7.444 \times 10^{-2} + j8.431\Omega$. The capacitance needed to tune the loop to resonance, from (7.12), is 377.5 pF. Lastly, from (7.13), Z'_{in} at resonance is equal to 955Ω.

7.3.1.2 Miniature Spiral

A second type of antenna that is employed in implantable biosensors, as listed in Table 7.2, consists of a printed multiturn spiral fabricated on a dielectric substrate. In general, the substrate material could be plastic, ceramic, or a semiconductor-like high-resistivity silicon. In addition, the spiral can be printed in different shapes and dimensions, depending on the requirements. The most common shapes are square, hexagonal, octagonal, and circular, as illustrated in Figure 7.8. The square spirals are the most popular because of the ease of layout. In this section, three simple and accurate expressions to predict the inductance of a spiral are presented [25]. For a given shape, an inductor/antenna is completely specified by the following:

N = number of turns

w = turn or strip conductor width

s = gap or spacing between turns

d_{out} = outer diameter of the spiral

d_{in} = inner diameter of the spiral

d_{avg} = average diameter of the spiral, equal to $0.5(d_{out} + d_{in})$

ρ = fill ratio, equal to $(d_{out} - d_{in})/(d_{out} + d_{in})$

t = turn or strip conductor thickness

Three expressions are described below. The first expression is based on a modification of an equation presented by Wheeler [26]. The second expression is derived from electromagnetic principles, by approximating the sides of the spirals as cur-

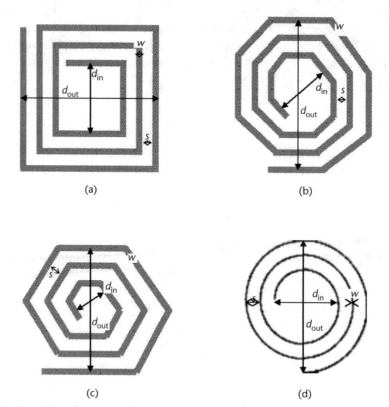

Figure 7.8 On-chip multiturn spiral inductor/antenna: (a) square, (b) octagonal, (c) hexagonal, and (d) circular. The strip and slot or gap widths are denoted as w and s, respectively. The inner and outer diameters of the spiral are denoted as d_{in} and d_{out}, respectively.

rent-strips. The third is a monomial expression derived from fitting several thousand inductors to a large database [25]. All three expressions are accurate, with errors typically ranging from 2% to 3%, and are excellent for use in design and synthesis.

The first expression is called a modified Wheeler (MW) expression [25]:

$$L_{MW} = K_1 \mu_0 \frac{N^2 d_{avg}}{1 + K_2 \rho} \tag{7.14}$$

where K_1 and K_2 are layout-dependent, and are presented in Table 7.3. Two inductors with identical d_{avg} but different ρ will have different inductance values. The circuit with a higher ρ will have a smaller inductance due to destructive mutual coupling effects.

Table 7.3 Coefficients for Modified Wheeler Expression

Layout	K_1	K_2
Square	2.34	2.75
Hexagon	2.33	3.82
Octagon	2.25	3.55

The second expression is based on current strip approximation (CSA) [25]:

$$L_{csa} = \frac{\mu_0 N^2 d_{avg} c_1}{2} \left[\ln\left(\frac{c_2}{\rho}\right) + c_3\rho + c_4\rho^2 \right]$$

(7.15)

where the coefficients C_i $(i = 1$ to $4)$ are layout-dependent, and are presented in Table 7.4. The accuracy of the above expression worsens as the ratio s/w increases. However, inductors are typically fabricated with $s \leq w$, because the inductor will occupy a small area on the wafer. Although making s smaller increases the parasitic interwinding capacitance, it is not a concern, since it is small compared to the underpass capacitance, as will be discussed later.

Example 7.5 As a numerical example, consider an implantable biosensor integrated with a square spiral chip antenna fabricated on a high-resistivity silicon substrate, having $s = 10$ μm, $w = 15$ μm, $N = 10$, $d_{in} = 520$ μm, and $d_{out} = 1,000$ μm. The calculated values for the inductances L_{MW} and L_{CSA} from (7.14) and (7.15) are 115.1 and 113.8 nH, respectively.

The third expression is called the data-fitted monomial (DFM) expression [25]:

$$L_{DFM} = \beta(d_{out})^{a_1} (w)^{a_2} (d_{avg})^{a_3} (N)^{a_4} (s)^{a_5}$$

(7.16)

where the coefficients a_i $(i = 1$ to $5)$ and the term β are layout-dependent, and are presented in Table 7.5. In addition, d_{out}, w, d_{avg}, and s are expressed in microns, and the calculated value of L_{DFM} is in nanohenrys, as illustrated in Example 7.6.

Example 7.6 For the square spiral chip antenna in Example 7.5, the inductance is calculated from (7.16) as

$$L_{DFM} = 1.62 \times 10^{-3} \times (1,000)^{-1.21} \times (15)^{-0.147} \times (760)^{2.40} \times (10)^{1.78} \times (10)^{-0.03}$$
$$= 117.6 \text{ nH}$$

Table 7.4 Coefficients for Expression Based on Current Strip Approximation

Layout	C_1	C_2	C_3	C_4
Square	1.27	2.07	0.18	0.13
Hexagon	1.09	2.23	0.00	0.17
Octagon	1.07	2.29	0.00	0.19
Circle	1.00	2.46	0.00	0.20

Table 7.5 Coefficients for Data-Fitted Monomial Expressions

Layout	β	a_1	a_2	a_3	a_4	a_5
Square	1.62×10^{-3}	−1.21	−0.147	2.40	1.78	−0.030
Hexagon	1.28×10^{-3}	−1.24	−0.174	2.47	1.77	−0.049
Octagon	1.33×10^{-3}	−1.21	−0.163	2.43	1.75	−0.049

Note that the calculated values of the inductance using the above three expressions are almost identical.

The miniature spiral antenna can be represented for the purpose of computer modeling by a lumped equivalent circuit model, as shown in Figure 7.9. The inductance L_s can be calculated by using any of the equations (7.14), (7.15), and (7.16). The series loss resistance R_L is given by [27]

$$R_L = \frac{l}{\left(\sigma w \delta \left(1 - e^{\frac{-t}{\delta}}\right)\right)} \quad \Omega \tag{7.17}$$

where:

$l = 4Nd_{avg}$ in meters

w = turn or strip conductor width in meters

t = turn or strip conductor thickness in meters

= skin depth, which is expressed as $\sqrt{\dfrac{2}{\omega \mu_0 \sigma}}$

$\omega = 2\pi f$

f = frequency in hertz

$\mu_0 = 4\pi \times 10^{-7}$ H/m

σ = conductivity of the strip conductor metal (for gold, $\sigma = 4.1 \times 10^7$ S/m)

The series loss resistance represents the energy lost due to the skin effect losses in the strip conductor. The series capacitance is mainly due to the capacitance between the spiral and the metal underpass required to connect the inner end of the spiral inductor to the external circuit, and is given by [27]

$$C_s = \frac{\varepsilon_0 \varepsilon_r N w^2}{t_{diel}} \tag{7.18}$$

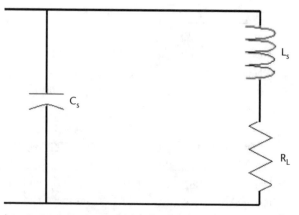

Figure 7.9 Lumped equivalent circuit model for the spiral antenna.

where $\varepsilon_0 = 8.854 \times 10^{-12}$ F/m, and ε_r is the relative permittivity of the dielectric between the spiral and the underpass. For silicon dioxide and silicon nitride, the values for ε_r range from 4 to 5 and 5.5 to 7.5, respectively. t_{diel} is the thickness of the dielectric between the spiral and the underpass. The effect of the interturn fringing capacitance is ignored, since the adjacent turns are almost equipotential.

Example 7.7   Consider the square spiral chip antenna discussed above. Assuming that the strip conductor is gold metallization with a thickness of 1 μm and the frequency of operation is 400 MHz, the series resistance from (7.17) is 56Ω. In addition, if we assume that the dielectric between the spiral and the underpass is 1 μm thick and has a relative permittivity of 5.0, then from (7.18), the series capacitance is approximately 100 fF.

7.3.2   Antennas for External Handheld Devices

In the previous section, it was shown that the inductive near field dominates in the vicinity of the implanted antenna. Thus, the most efficient method of receiving telemetry data from the implanted sensor is through inductive or magnetic coupling. Based on this principle, several researchers have demonstrated antennas for telemetry reception from implantable biosensors [7, 12, 28, 29]. These antennas include a planar spiral, disk coil, solenoid coil, and printed multiturn loop, and Table 7.6 summarizes their key features. These antennas can be considered as variants of the basic loop antenna. In this section, we will present the design guidelines for the impedance matching of a simple loop of wire or printed circuit board trace to a receiver, using a tapped capacitor circuit topology that is tuned to resonate at a desired frequency.

7.3.2.1   Loop Antenna

Typically, the loop antenna in the handheld device would have a diameter greater than the diameter of the antenna integrated with the implant. However, the antenna circumference will still be less than approximately 0.2λ, where λ is the free-space wavelength. Therefore, the uniform current distribution approximation is still valid, and the expressions for the radiation resistance, loss resistance, and inductance, given by (7.7), (7.8), (7.9), and (7.10), respectively, also hold good in this case. Figure 7.10(a, b) presents a single turn loop antenna with a tapped capacitor-matching network, and the equivalent circuit model schematic, respectively. In the equivalent circuit model, R_r, R_L, and L_A represent the radiation resistance, loss

Table 7.6 Summary of the State-of-the-Art Antennas for External Handheld Devices in RF Biotelemetry Systems

Type of Antenna	Antenna Diameter (mm)	Reference
Planar spiral	80	Von Arx and Najafi [7]
Disk coils	90	Hamici, Itti, and Champier [28]
Solenoid coils	100 to 150	Troyk and Edgington [29]
Printed multiturn loop	51	Simons, Hall and Miranda [12]

resistance, and the inductance of the loop antenna, respectively. The loop antenna input impedance Z_{Loop} is given by

$$Z_{\text{Loop}} = R_t + j\omega L_A \tag{7.19}$$

where

$$R_t = R_r + R_L \quad \Omega \tag{7.20}$$

The Z_{Loop} has to be matched to the impedance of the receiver for maximum power transfer. Integrating a pair of capacitances C_1 and C_2 and tapping the signal, as shown in Figure 7.10, accomplish this task. The capacitances C_1 and C_2 form a matching network that transforms the impedance and resonates the antenna. The admittance Y_{in} looking into the capacitor tap is given by

$$Y_{\text{in}} = G_{\text{in}} + jB_{\text{in}} \tag{7.21}$$

where

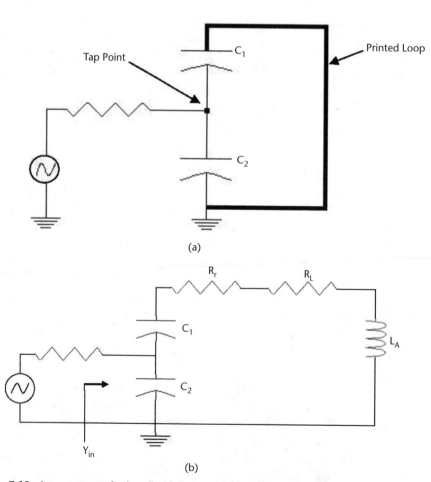

Figure 7.10 Loop antenna for handheld device: (a) physical representation of the loop and the tapped capacitor impedance matching network, and (b) equivalent circuit model.

$$G_{in} = \frac{R_t}{R_t^2 + \left(\omega L_A - \dfrac{1}{\omega C_1}\right)^2} \tag{7.22}$$

$$B_{in} = \left(\omega C_2 - \frac{\omega L_A - \dfrac{1}{\omega C_1}}{R_t^2 + \left(\omega L_A - \dfrac{1}{wC_1}\right)^2}\right) \tag{7.23}$$

At resonance, the reciprocal of the real part of Y_{in} gives the input impedance Z_{in}, which is expressed as

$$Z_{in} = \frac{R_t^2 + \left(\omega_0 L_A - \dfrac{1}{\omega_0 C_1}\right)^2}{R_t} \quad \Omega \tag{7.24}$$

Equation (7.24) is solved for the capacitance C_1, which is given by

$$C_1 = \frac{1}{\omega_0 \left(\omega_0 L_A - \sqrt{Z_{in} R_t - R_t^2}\right)} \tag{7.25}$$

where $\omega_0 = 2\pi f_0$ and f_0 is the resonance frequency.

Setting the imaginary part of Y_{in} at resonance to zero and solving for the capacitance C_2 gives

$$C_2 = \frac{L_A - \dfrac{1}{\omega_0^2 C_1}}{R_t^2 + \left(\omega_0 L_A - \dfrac{1}{\omega_0 C_1}\right)^2} \quad F \tag{7.26}$$

The resonance frequency f_0 is related to the matching network and antenna parameters through the expression

$$f_0 = \frac{1}{2\pi \sqrt{L_A \left(\dfrac{C_1 C_2}{C_1 + C_2}\right)}} \quad Hz \tag{7.27}$$

By substituting (7.27) in (7.24), we obtain the Z_{in} at resonance, which is given by

$$Z_{in} = R_t + \frac{L_A}{R_t \left(\dfrac{C_1 C_2}{C_1 + C_2}\right)} \left(\frac{C_1}{C_1 + C_2}\right)^2 \quad \Omega \tag{7.28}$$

The following example illustrates the application of (7.25) and (7.26) in the design of the impedance matching network for the antenna in the handheld device.

Example 7.8   Consider a circular printed loop antenna with a mean diameter of 4 cm fabricated on a 1-oz copper-clad FR-4 board material (ε_r = 4.8) that is approximately 1.6 mm thick. The width w and thickness t of the copper trace that forms the loop are 2 mm and 34 μm, respectively. The antenna operates at 300 MHz, and is matched to a receiver with an input impedance of 1 kΩ.

From (7.7) and (7.8), the R_r and R_L are 0.05Ω and 0.141Ω, respectively. Since (7.9) for the antenna inductance assumes that the loop is constructed from a wire of radius b, the copper trace of width w is transformed to a wire with an electrical equivalent radius of 0.25w [23]. Then, from (7.9), L_A is 94.71 nH. The capacitances C_1 and C_2 from (7.25) and (7.26) are 3.22 and 38.48 pF, respectively. Notice that the capacitance $C_2 >> C_1$.

7.4 Space, Military, and Civilian Applications

In general, there are two types of body sensors—sensors that are implanted inside the body, and sensors that are in close proximity to the body, but worn on the outside. Possible applications for these sensors include spaceflight, military applications, medical applications, and in the home. In this section, examples of implanted and body-worn sensors are given for each application.

7.4.1 Sensors for Space Environment

The primary motivation for the application of sensors in space is for a greater understanding of the impact of spaceflight on living systems [30]. Research subjects, such as small animal models, have to be fully untended and unrestrained for long periods to fully reflect the impact of microgravity and spaceflight on their behavior and physiology [30]. These requirements preclude the use of hard-wired instrumentation and data acquisition systems, and favor innovative implantable sensors with miniaturized biotelemetry systems.

7.4.1.1 Dual-Channel Electrocardiogram/Temperature Biotelemeter

The biotelemeter [30] is totally implantable and is approximately 20 cc in volume, with a transmission lifetime from 6 to 9 months. It digitally encodes data and transmits "ON-OFF" keyed bursts of 455 KHz RF at rates equivalent to 4,096 baud. The receiver is lightweight and portable, and can acquire signals at a distance ranging from 30 to 60 cm. This work was later extended to include a third data channel for pH monitoring [31, 32]. Recently, a wearable multiparameter ambulatory physiologic monitoring system for space and terrestrial applications has been presented [33].

7.4.1.2 Sensor Pills for Physiological Monitoring

A miniature biotelemetry unit resembling a large pill has been proposed for use in physiological monitoring of the gastrointestinal tract [34]. The proposed pill would carry sensors to provide indication of the presence of blood, bacteria, and chemicals of interest. Furthermore, it will not contain any batteries, and would be inductively powered when interrogated by the hand-held transceiver, as discussed in Section 7.2. The length and diameter of the sensor pill are typically 0.5 and 2 cm, respectively. The pill can be swallowed, and would pass through the gastrointestinal tract in approximately 24 hours.

7.4.1.3 Sensor Patch for Physiological Monitoring

A wearable sensor patch (3 × 3 cm) has been proposed that will contain sensors and electronic circuits for measuring the temperature, heart rate, blood pressure, and possibly other physiological parameters [35]. As in the previous example, the sensor patch will not carry any batteries, and would be inductively powered when interrogated by the handheld device.

7.4.2 Battlefield Sensors

This type of sensor is to provide combat casualty care, and is targeted towards soldiers and support personnel in battlefields. Astronauts who are working inside spacecraft can also wear it.

7.4.2.1 Smart Shirt Technology

The smart shirt uses optical fibers and special sensors to detect bullet wounds, and monitors the body's vital signs, including heart rate, respiration rate, electrocardiogram (EKG), body temperature, and pulse oximetry during combat conditions [36]. In addition, by plugging a microphone into the smart shirt, voice can be recorded.

7.4.3 Sensors in Hospitals and Smart Homes

The possible reasons for using wireless sensors in hospitals and in smart homes are as follows [37]:

1. They can help patients comply with doctors' orders, enabling remote caregiving and detection of the early signs of disease its progress.
2. They can help in the diagnosis of neurological disorders, such as Parkinson's and Alzheimer's diseases. Parkinson's can be diagnosed only through behavioral change, principally changes in gait monitored through the use of a motion sensor.
3. They can monitor transient or infrequent events, such as a sudden slowing of the heart rate that leads to fainting spells that can last less than a minute and go unnoticed by the patient.
4. They can serve as a device to warn of any symptoms of heart failure.

A variety of body sensors for monitoring the above conditions are required, all with the capability to communicate using wireless technology.

If these body sensors are to be networked and operated seamlessly, then a personal mobile hub would be required [38]. By supporting multiple wireless protocols, some short-range and some wide-area, the hub will make available the power of the Internet to body-worn sensors.

References

[1]   http://www.nasa.gov/missions/solarsystem/explore_main.html.

[2]   Special Issue on Electromagnetic-Wave Interactions with Biological Systems, *IEEE Trans. on Microwave Theory and Techniques*, Vol. 32, No. 8, August 1984.

[3]   Special Issue on Phased Arrays for Hyperthermia Treatment of Cancer, *IEEE Trans. on Microwave Theory and Techniques*, Vol. 34, No. 5, May 1986.

[4]   Special Issue on Medical Application and Biological Effects of RF/Microwaves (Part II), *IEEE Trans. on Microwave Theory and Techniques*, Vol. 44, No. 10, October 1996.

[5]   Mini-Special Issue on RF/Microwave Applications in Medicine (Part I) and Special Issue on Medical Application and Biological Effects of RF/Microwaves (Part II), *IEEE Trans. on Microwave Theory and Techniques*, Vol. 48, No. 11, November 2000.

[6]   Special Issue on Medical Applications and Biological Effects of RF/Microwaves (Part II), *IEEE Trans. on Microwave Theory and Techniques*, Vol. 52, No. 8, August 2004.

[7]   Von Arx, J. A., and K. Najafi, "On-Chip Coils with Integrated Cores for Remote Inductive Powering of Integrated Microsystems," *1997 Int. Conference on Solid-State Sensors and Actuators, TRANSDUCERS '97*, Chicago, IL, June 16–19, 1997, pp. 999–1002.

[8]   Eggers, T., et al., "Advanced Hybrid Integrated Low-Power Telemetric Pressure Monitoring System for Biomedical Applications," *13th Annual Int. Conference on Microelectromechanical Systems Digest*, Miyazaki, Japan, January 23–27, 2000, pp. 329–334.

[9]   Ullerich, S., et al., "Micro-coils for an Advanced System for Measuring Intraocular Pressure," *Tech. Digest. 1st Annual Int. IEEE-EMBS Special Topic Conference on Micro- technologies in Medicine & Biology*, Lyon, France, October 12–14, 2000, pp. 470–474.

[10] Mokwa, W., and U. Schnakenberg, "Micro-Transponder Systems for Medical Applications," *IEEE Trans. on Instrumentation and Measurement*, Vol. 50, No. 6, December 2001, pp. 1551–1555.

[11] Simons, R. N., D. G. Hall, and F. A. Miranda, "RF Telemetry Systems for an Implantable Bio-MEMS Sensor," *2004 IEEE MTT-S Int. Microwave Symp. Digest*, Vol. 3, Fort Worth, TX, June 6–11, 2004, pp. 1433–1436.

[12] Simons, R. N., D. G. Hall, and F. A. Miranda, "Printed Multi-Turn Loop Antenna for RF Bio-Telemetry," *2004 IEEE Int. Symp. Antennas and Propagation and USNC/URSI National Radio Science Meet. Digest*, Vol. 2, Monterey, CA, June 20–26, 2004, pp. 1339–1342.

[13] Miranda, F. A., R. N. Simons, and D. G. Hall, "Validation of Radio Frequency Telemetry Concept in the Presence of Biological Tissue-Like Stratified Media," *2004 IEEE Int. Symp. Antennas and Propagation and USNC/URSI National Radio Science Meet. Digest*, Vol. 2, Monterey, CA, June 20–26, 2004, pp. 1335–1338.

[14] Simons, R. N., D. G. Hall, and F. A. Miranda, "Spiral Chip Implantable Radiator and Printed Loop External Receptor for RF Telemetry in Bio-Sensor Systems," *Proc. 2004 IEEE Radio and Wireless Conference (RAWCON)*, Atlanta, GA, September 19–22, 2004, pp. 203–206.

[15] Simons, R. N., and F. A. Miranda, "Radiation Characteristics of Miniature Silicon Square Spiral Chip Antenna for Implantable Bio-MEMS Sensors," *2005 IEEE Int. Symp. Antennas and Propagation and USNC/URSI National Radio Science Meet. Digest*, Vol. IB, Washington, July 3–8, 2005, pp. 836–839.

[16] Simons, R. N., and F. A. Miranda, "Radio Frequency Telemetry System for Sensors and Actuators," U.S. Patent No. 6667725, December 23, 2003.

[17] Yue, C. P., and S. S. Wong, "On-Chip Spiral Inductors with Patterned Ground Shields for Si-Based RF IC's," *IEEE J. Solid-State Circuits*, Vol. 33, No. 5, May 1998, pp. 743–752.

[18] Chang, C. A., et al., "Characterization of Spiral Inductors with Patterned Floating Structures," *IEEE Trans. on Microwave Theory and Techniques*, Vol. 52, No. 5, May 2004, pp. 1375–1381.

[19] Ahmadian, M., et al., "Miniature Transmitter for Implantable Micro Systems," *Proc. of the 25th Annual Int. Conference of the IEEE Engineering in Medicine and Biology Society*, Vol. 4, Cancun, Mexico, September 17–21, 2003, pp. 3028–3031.

[20] Mokwa, W., "Ophthalmic Implants," *Proc. 2nd IEEE Int. Conference on Sensors*, Vol. 2, Toronto, Canada, October 22–24, 2003, pp. 980–986.

[21] Van Schuylenbergh, K., et al., "Monitoring Orthopedic Implants Using Active Telemetry," *Proc. of the 14th Annual Int. Conference of the IEEE Engineering in Medicine and Biology Society, EMBS-92*, Vol. 6, Paris, France, October 29–November 1, 1992, pp. 2672–2673.

[22] Akin, T., et al., "A Wireless Implantable Multichannel Digital Neural Recording System for a Micromachined Sieve Electrode," *IEEE J. Solid-State Circuits*, Vol. 33, No. 1, January 1998, pp. 109–118.

[23] Balanis, C. A., *Antenna Theory Analysis and Design*, 2nd ed., Chap. 5 and 9, Table 9.3, New York: John Wiley & Sons, 1997.

[24] Ramo, S., J. R. Whinnery, and T. V. Duzer, *Fields and Waves in Communication Electronics*, 3rd ed., New York: John Wiley & Sons, 1994, p. 81.

[25] Mohan, S. S., et al., "Simple Accurate Expressions for Planar Spiral Inductances," *IEEE J. Solid-State Circuits*, Vol. 34, No. 10, October 1999, pp. 1419–1424.

[26] Wheeler, H. A., "Simple Inductance Formulas for Radio Coils," *IRE Proc.*, Vol. 16, No. 10, October 1928, pp. 1398–1400.

[27] Hershenson, M. D. M., et al., "Optimization of Inductor Circuits via Geometric Programming," *Proc. 36th Design Automation Conference*, New Orleans, LA, June 21–25, 1999.

[28] Hamici, Z., R. Itti, and J. Champier, "A High-Efficiency Biotelemetry System for Implanted Electronic Device," *IEEE Engineering in Medicine and Biology Society 17th Annual Conference*, Vol. 2, Montreal, Canada, September 20–23, 1995, pp. 1649–1650.

[29] Troyk, P. R., and M. Edgington, "Inductive Links and Drivers for Remotely-Powered Telemetry Systems," *IEEE Antennas and Propagation Society Inter. Symp. Digest*, Vol. 1, Salt Lake City, UT, July 16–21, 2000, pp. 60–62.

[30] Hines, J. W., "Medical and Surgical Applications of Space Biosensor Technology," *Acta Astronautica (Special Issue on Benefits of Space for Humanity)*, Vol. 38, Nos. 4–8, February–April, 1996, pp. 261–267.

[31] Hines, J. W., et al., "Telemetric Sensors for the Space Life Sciences," *Proc. 18th Annual Int. Conference of the IEEE Engineering in Medicine and Biology Society*, Vol. 1, Amsterdam, Holland, October 31–November 3, 1996, pp. 30–31.

[32] Hines, J. W., et al., "Space Biosensor Systems: Implications for Technology Transfer," *Proc. 19th Annual Int. Conference of the IEEE Engineering in Medicine and Biology Society*, Vol. 2, Chicago, IL, October 30–November 2, 1997, pp. 740–743.

[33] Mundt, C. W., et al., "A Multiparameter Wearable Physiologic Monitoring System for Space and Terrestrial Applications," *IEEE Trans. on Information Technology in Biomedicine*, Vol. 9, No. 3, September 2005, pp. 382–391.

[34] http://www.nasatech.com/Briefs/Feb00/NPO20652.html (Improved Sensor Pills for Physiological Monitoring).

[35] http://www.nasatech.com/Briefs/Feb00/NPO20651.html (Wearable Sensor Patches for Physiological Monitoring).

[36] Park, S., and S. Jayaraman, "Enhancing the Quality of Life Through Wearable Technology," *IEEE Engineering in Medicine and Biology Magazine*, Vol. 22, No. 3, May/June 2003, pp. 41–48.

[37] Ross, P. E., "Managing Care Through the Air," *IEEE Spectrum*, Vol. 41, No. 12, December 2004, pp. 26–31.

[38] Husemann, D., et al., "Personal Mobile Hub," *Proc. 8th Int. Symp. Wearable Computers (ISWC'04)*, Vol. 1, October 31–November 3, 2004, pp. 85–91.

CHAPTER 8

Antennas and Propagation for Telemedicine and Telecare: On-Body Systems

William Scanlon and Noel Evans

This chapter examines antennas and propagation considerations for wearable telemedicine and telecare systems at frequencies above 100 MHz. At these frequencies, the main propagating mode is a radiating wave, even for devices that are located in close proximity to each other on the body surface. The chapter first considers both traditional and emerging telemedicine and telecare applications before detailing specific antenna-body interaction and channel effects. The chapter concludes by briefly discussing the important issue of radio frequency interference in telemedicine.

8.1 Telemedicine and Telecare Applications

Telemedicine is often rather bluntly defined as *medicine at a distance*—a service with the potential to have a favorable impact on the access, quality, and cost of health care [1]. Wireless communication is therefore an important technology enabler for telemedicine, offering the potential for both patient mobility and ubiquity of service. In particular, radio communication has been, and will continue to be, the dominant technique for wireless telemedicine. Although the alternatives to radio have been used for niche applications, they are either limited to line-of-sight environments (e.g., diffuse IR [2]), or suffer from extremely low bandwidths (e.g., ultrasonic [3]). A good introduction and overview of wireless telemedicine can be found in [4]. The definition of telecare is wider than that of telemedicine, involving the delivery of health and social care applications, including those where medical supervision is not required, to individuals within the home or wider community [5].

Personal telemedicine can be defined as those systems that are aimed at providing health care services for individual patients on a short-range wireless or a full-mobility basis. It is anticipated that, through the continuing development of personal communications technology, such as WiFi-based (IEEE 802.11 [6]) WLANs and third generation mobile telephony (e.g., UMTS/CDMA2000), personal telemedicine will become a major mode of health care delivery. There are a number of difficulties, including patient and clinician acceptability, privacy, security, and the need for supporting infrastructure. An effective personal telemedicine system also needs comfortable, lightweight, wearable communications equipment. While the antennas and propagation aspects of telemedicine technology have similarities with other body-centric applications, transmitted power levels are generally much lower and antenna efficiencies reduced because of battery and packaging constraints. Furthermore, patient-worn devices should ideally be extremely low-cost and potentially disposable. These factors mean that the antennas and propagation aspects of wearable telemedicine systems are very important for both academic research and engineering practice.

The wider definition of telemedicine covers all uses of communication technology, including data links with diverse devices, such as sensors, actuators (e.g., bladder or muscle stimulators), prostheses, and controllers/processors. Furthermore, this diversity will continue to grow as the market and technologies develop. However, patient monitoring [7] remains the most pervasive application of wireless communications in telemedicine, with an increasing need for the remote monitoring of vital signs, in applications such as the care of day-surgery patients recovering at home or in emergency trauma signaling from accident sites. Wireless patient monitoring is sometimes referred to as *medical telemetry*. While the term *biotelemetry* is similarly defined, it concerns both human and animal signal acquisition. Figure 8.1 illustrates the concept of a fully connected patient-centric wearable telemedicine system, highlighting both existing and emerging technologies. In this approach, a network of low-power wearable devices (sensors or actuators, surface-worn or implanted) is established using short-range UHF radio. However, these devices may not have suf-

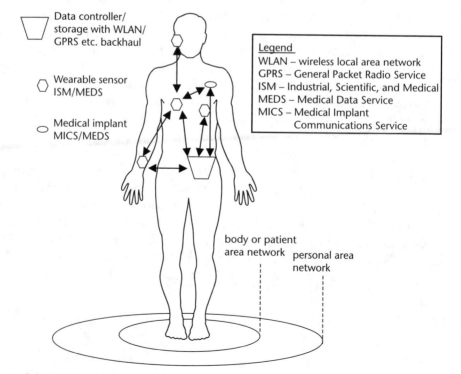

Figure 8.1 Patient-centric wearable telemedicine system.

ficient data processing, energy, or memory resources to fully realize their function. Thus, a more substantial controller or base station device may also be part of the wearable network. The controller could be a multifunction device, such as a personal digital assistant (PDA), cell phone [8], or a fully proprietary telemedicine unit. Regardless of its implementation, the controller could introduce a range of backhaul communication technologies, including WLAN, digital enhanced cordless telecommunication (DECT), or general packet radio service (GPRS), providing important local feedback to both patients and caregivers. Figure 8.1 also indicates the extent of a body area network (BAN) that would incorporate all devices worn on, or implanted in, the body. Similarly, a personal area network (PAN) incorporates all wearable devices, but its range is extended to include nearby (e.g., from 3m to 10m) devices, such as a bedside monitor, notebook PC, and so forth.

8.1.1   Physiological Signals for Patient Monitoring

Remote patient monitoring has had a relatively long history. Since the introduction of *electrocardiography* by Einthoven [9] at the turn of this century, its use, importance, and acceptance have rapidly expanded. In electrocardiography, a noninvasive, simple, inexpensive, and reproducible procedure allows the EKG to be recorded, which provides sufficient cardiac information to permit an initial diagnosis. Thus, the ability to present this information from a remote location to be analyzed by a specialist offers a significant advantage where the levels of expertise are not available to evaluate the patient onsite. Indeed, Einthoven's initial work [9] described the transmission of EKG information over telephone wires. Although at the time this was not labeled as a telemedicine system, or more specifically *telecardiology*, the principles of remote EKG transmission were ably demonstrated. The underlying fundamentals of the process have changed little over the subsequent century, and the EKG remains one of the most important physiological parameters for patient monitoring. Other basic parameters that may be transmitted range from simple heart rate and body temperature to full-bandwidth electroencephalogram (EEG) waveforms and blood glucose measurements. Table 8.1 lists some of the more common physiological parameters, and details their sampling rate and typical baseline information bandwidth requirements, ignoring framing, error coding, encryption, and other protocol overheads. More detailed information on physiological parameters can be found in [10]. Table 8.1 illustrates the extent of bandwidth requirements for remote patient monitoring, with most other signals falling within the range given.

8.1.2   Technologies for Ward-Based Systems

Existing hospital-based medical telemetry systems are usually for EKG or oxygen saturation-pulse oximetry (S_pO_2) monitoring. These are mostly simplex systems operated on a secondary basis in land mobile spectrum allocations from 450 to 470 MHz. This spectrum is shared with other users, and the systems are subject to interference causing regular loss of data [11]. In 2000, the FCC established a wireless medical telemetry service (WMTS) with a total of 14 MHz of spectrum, including 608 to 614 MHz, 1,395 to 1,400 MHz, and 1,429 to 1,432 MHz [12]. This service

Table 8.1 Common Signal Parameters in Patient Monitoring

Signal	Typical Amplitude, Range, and Bandwidth	Clinical Resolution	Bits/Sample	Typical Sample Rate (Hz)	Bandwidth (Kbps)
Body temperature	30°C–45°C (essentially a dc signal)	0.1°C	8	0.01	~0
Fluid output	0–1,000 mL/h	1 mL/h	10	0.1	~0
Fluid intake	0–100 drops/min	1 drop/min	8	1	~0
Respiration	12–40 breaths/min	1 breath/min	8	0.01	~0
S_pO_2	70%–100%	1%	8–10	100	1.0
Blood pressure	Systolic 0–250 mm Hg Diastolic 0–200 mm Hg dc–60 Hz	1 mm Hg	9	120	1.0
EKG	1 mV p-p 0.05–100 Hz nominal 0–200 bpm Typically 3 leads	10 μV	8–12	250	9.0
EEG	10–100 μV p-p (up to 64 channels) 0.5–60 Hz	1 μV	12	200	153.6

is primarily designed for wireless patient monitoring in hospitals (or equivalent health care facilities), and there are strict rules governing transmitter output power, spurious emissions, and interference with other services. WMTS equipment is not licensed for use in ambulances or anywhere outside the hospital. While WMTS has the advantage of being a primary service in the spectrum allocated, restrictions on data content (e.g., no voice or video), issues of vendor interoperability, and the lack of international harmonization, especially within Europe, still restrict its role in wireless telemedicine.

For valid reasons, including the presence of a global agreement on unlicensed operation and high bandwidth potential, the 2.45 GHz ISM band is becoming extremely popular for low-power, short-range to midrange device applications, including personal and local area networks. In contrast to the use of WMTS, the use of general purpose communications standards, such as 802.11 WLAN and Bluetooth, for telemedicine in this band offers fewer restrictions and lower costs. However, since these technologies are not usually formally coordinated, their performance depends on many uncontrollable factors. For example, in the most popular version of 802.11 WLAN, there are only three nonoverlapping channels. Performance may be poor due to high interference levels in those installations without proper frequency planning. Nonetheless, there are some good examples of using WLAN for patient monitoring in the literature, including [13]. Note that the restriction on mobile cellular systems in clinical areas means that cellular data services such as GPRS cannot be used.

Power consumption remains a major concern for wearable devices in ward-based systems. In particular, only the shorter range ISM (i.e., 433 to 868 MHz in Europe, and 315 to 915 MHz in North America) and lower power Bluetooth devices (i.e., 0 dBm, class 3) have sufficiently small energy requirements to offer a reasonable trade-off between size, weight, and operational lifetime. Some applications may also use communications across bands to reduce power. For example, in [14], a passive 2.45 GHz receiver is used in conjunction with a 418-MHz SRD transmitter to create a low-power telecommand system, in which patient information is available on demand.

8.1.3   Technologies for Home-Based and Full-Mobility Systems

It is generally considered beneficial for many health care activities to take place within the patient's home rather than in the clinical environment. As with ward-based systems, there are wide ranges of suitable technologies available for wireless telemedicine in the home. Even though WMTS systems are not available, both short-range [e.g., short-range device (SRD)] and midrange (e.g., Bluetooth, WiFi, and DECT) ISM devices are suitable replacements. DECT modems are particularly suitable for home-based monitoring, since the frequencies employed (1,880 to 1,900 MHz in Europe, and 2.4 GHz in the United States) mean that the patient antennas are more efficient. The channel selection mechanism used automatically minimizes interference between units and other users of the spectrum. Furthermore, the use of mobile cellular technologies to introduce full mobility to a personal telemedicine system is an obvious step. A fully proprietary system can easily incorporate a cellular data modem using GPRS or similar technology. Alternatively, a standard cellular handset may be used to provide the backhaul with either a direct, hardwired, or wireless (e.g., Bluetooth) connection to the wearable telemedicine unit [15]. If full mobility is not needed, then a public switched telephone network (PSTN) or broadband asymmetric digital subscriber line (ADSL) modem can provide an extremely reliable backhaul to the wider telemedicine service.

8.1.4   Emerging Technologies and Novel Applications

Irrespective of whether the telemedicine system is within the hospital or home environment, the short-range link between the patient and base station must be implemented using either low-bandwidth, 300-to-900-MHz unlicensed short-range devices, or, where higher data rates are needed, using UWB, WPAN, or WLAN technology. The low-bandwidth systems are typically proprietary, extremely low power, and low cost; and the design of the patient's antenna is often more difficult due to the longer wavelengths involved. Current developments in this field include a proposal to establish a Medical Data Service (MEDS) in the 401 to 402 MHz and 405- to 406-MHz bands [16] to complement the existing Medical Implant Communications Service (MICS), as will be described in Chapter 9. The proposed MEDS service would operate on a secondary, noninterfering basis with existing meteorological aids and related satellite services. A key aspect of MEDS is that it allows for ultralow-power, transmit-only devices with an EIRP restriction of 250 nW, at a maximum duty cycle of 0.1% and a maximum communications bandwidth of 100 kHz. For devices that employ suitable listen-before-talk procedures, the EIRP limit is increased to 25 μW. The advantage of MEDS devices is that they will operate in a less noisy environment than do other ISM-based SRD devices, and even with current technology, it is possible to engineer low-cost disposable transmit-only sensors for integration within a WBAN, as shown in Figure 8.1. It is likely that emerging telemedicine applications will employ this or a similar WBAN configuration. If so, the requirements for antennas will be in common with the requirements in other developing applications, such as wearable computing [17]. Examples of these requirements include: high efficiency (i.e., reduced power loss in body tissue and increased safety), multiband or wideband performance, low profile and physically small design, reduced sensitivity to body proximity, and conformability.

Patient localization is emerging as an important incidental benefit of wireless telemedicine. Depending on the application, localization can be used either in response to a medical incident or as part of the monitoring system itself (e.g., providing information on levels of activity). Although the received signal strength can be used to obtain a relatively coarse position estimate using standard WLAN equipment in indoor environments [18], better performance can be obtained with proprietary spread spectrum–based signal processing [19]. Regardless of the approach adopted, radio-based personal telemedicine equipment can usually be configured to provide basic localization information. Outdoor localization can be achieved using a variety of methods, including cellular- or GPS-based systems [20].

8.1.5 Wireless Telemedicine Link Design

Like many other wearable communication applications, personal telemedicine link design involves the estimation of many unknown factors. However, it is still helpful to be able to identify a baseline link budget. Consider a source of average power P_T, radiating energy equally in all directions (isotropically). The energy spreads out spherically as it travels away from the source, so that at a distance d, the power density in the wave, P_{Di}, the power per unit area of the wavefront (in W/m^2), is:

$$P_{Di} = \frac{P_T}{4\pi d^2} \tag{8.1}$$

All practical antennas have directional characteristics; that is, they radiate more power in some directions at the expense of others. Body reflections can aid this process (see Section 8.2). With G_T being the maximum directivity gain of the transmitting antenna, relative to isotropic, then the power density along the direction of maximum radiation is:

$$P_D = P_{Di} \cdot G_T = \frac{P_T G_T}{4\pi d^2} \tag{8.2}$$

The power P_R delivered to a matched receiver by a corresponding receiving antenna with maximum directivity gain G_R can be shown to be:

$$P_R = P_T G_T G_R \left[\frac{\lambda}{4\pi d}\right]^2 \tag{8.3}$$

where λ is the wavelength of the wave being radiated.

Note that $P_R \propto (1/d^n)$, with $n = 2$ representing the power decay index in free space. It is rare for n to be precisely 2 in practice. In cluttered indoor environments, such as a hospital ward or a patient's home, it can rise to as much as 6. Enhancement to $n < 2$ is possible in smooth guiding environments, such as hospital corridors.

Equation (8.3) is the fundamental link power budget for free-space transmission, often known as Friis' free-space equation. It may be expressed in decibel form to give the path loss for the link, initially using isotropic antennas with $G_T = G_R = 1$:

$$P_L(\mathrm{dB}) = 10\log(P_R/P_T) = 20\log d(\mathrm{m}) + 20\log f(\mathrm{MHz}) - 27.6 \qquad (8.4)$$

Take the case of a body-worn 610-MHz WMTS device. The FCC rules for WMTS state that the maximum electric field strength, $E_{\max}$, at a distance of 3m from the transmitter, should not exceed 200 mV/m in this band. This field strength equates to a power density of 0.106 mW/m² ($P_{\mathrm{Dmax}} = E_{\max}^2/Z_0$, where Z_0 is the free-space wave impedance equal to 377Ω). Using (8.2), the maximum equivalent isotropic radiated power (EIRP) for a WMTS transmitter in the 610-MHz band is 12 mW (+10.8 dBm). However, to reduce battery size and weight, body-worn WMTS transmitters typically operate with a transmit power of 1 mW (0 dBm). Consider such a transmitter operating in a hospital over a 50-m unobstructed (LOS) path to a companion receiver. The path loss (8.4) is approximately 62 dB. A link power budget equation may be set up to give the power delivered to the receiver's input port:

$$P_{\mathrm{Rx-in}}(\mathrm{dBm}) = P_{\mathrm{Tx-out}}(\mathrm{dBm}) - P_L(\mathrm{dB}) + G_R(\mathrm{dBi}) + G_T(\mathrm{dBi}) \qquad (8.5)$$

For a modest receiving antenna, such as a quarter-wavelength whip antenna mounted on a small ground plane, a typical directive gain of approximately +2 dBi (gain relative to an isotrope) may be expected. The transmit antenna is usually integrated within the body-worn device, and it is unlikely that its maximum gain is more than −2 dBi, taking into account a nominal +2 dBi antenna gain and an estimated 4 dB of tissue losses at 600 MHz. Using (8.5), the maximum receiver input power is −62 dBm, well above the typical noise floor of −100 dBm in a 1.5-MHz channel. It must be emphasized that this simple, and wholly optimistic, model does not take into account multipath fading (approximately 30 dB maximum, effectively adding to path loss), induced by the host building's characteristics, and blocking by body tissue and other radio-opaque obstructions, such as walls and partitions. These are the most important factors in determining practical performance and link reliability.

8.2   Antennas and Human Body Interaction in Personal Telemedicine

The proximity of the human body affects the radiation patterns, feedpoint impedance, and efficiency of body-worn antennas in telemedicine applications. Such influence may render a marginal antenna (in terms of narrow bandwidth or efficiency) ineffective. The resultant system performance is difficult to predict, since such effects vary not only with frequency but also with antenna construction and feedpoint, body position, and tissue composition.

Until the late 1980s, little experimental or analytical data on antenna-body interaction for medical telemetry, which predates the concept of telemedicine, had appeared in the literature [21]. King et al. [22] presented an approximate analytical solution for implantable antennas using a three-layered model, while Neukomm [23] gave experimental results for body surface mounted telemeters. Amlaner [24] reviewed antennas for biotelemetry and recommended the use of whip, dustcore (ferrite), and simple loop antennas for transmission (body-worn use). More recently, research effort has increased in parallel with developments in the wider area of body-centric communications, with work in the early 1990s on antenna-body interaction measurements [25] and modeling [26, 27].

A constraining factor for radio-based personal telemedicine is the required size and effectiveness of body-worn antennas. The specification for body-worn antennas differs from that of base-station antennas. The body-worn unit is always required to be as small as possible. Other onboard hardware may be miniaturized, but the antenna size (maximum dimension) cannot normally be reduced below $\lambda/4$ while still retaining reasonable radiation characteristics. While this is readily achievable at 2.45 GHz with a 3-cm antenna, practical antennas will be electrically small at lower frequencies. Conversely, body losses are greater at higher frequencies, particularly for conformal or other low-profile antennas. Considering these factors, the requirements for body-worn antennas in personal telemedicine applications are as follows.

- They must be unobtrusive in size and construction.
- They must have reduced radiation pattern fragmentation. This effectively means an omnidirectional azimuthal radiation pattern, to allow links to off-body receivers. For on-body WBAN applications, additional directivity is needed in directions tangential to the body surface, certainly for operating frequencies above 1 GHz.
- They must be suitably matched to the RF transmitter for maximum power transfer, taking into account the potential variability in feedpoint impedance caused by small changes in antenna-body separation during natural movements.
- They must postively utilize the electromagnetic effects of human body proximity.
- They must be physically robust, although not necessarily rigid.

It is generally held that the negative effects of body proximity increase as antenna-body separation decreases, with the difference between placing a UHF antenna in a shirt pocket and a jacket pocket often being cited as giving an approximately 4 dB reduction in gain [28]. Over the frequency range of interest (100 MHz to 6 GHz), the effective conductivity, σ_{eff}, of skin, for example, increases from 0.5 to 4.0 Sm^{-1}, while the relative permittivity, ε_r, falls from 70 to 36. The equivalent magnitude of the wave impedance of skin is therefore 35Ω at 100 MHz, rising to 61Ω at 6 GHz. When a uniform electromagnetic wave intercepts the human body, this relatively low wave impedance reduces the close proximity electric field, while increasing the magnetic field. At a separation of $\lambda/4$, the body impedance is transformed and a high impedance is presented to the incoming wave, creating an increased electric field and a reduced magnetic field. At lower frequencies (<1 GHz), a deep null appears in the antenna's radiation pattern in the through-body direction, due to power absorption in the tissue, affecting both electric and magnetic fields. Mumford et al. [29] illustrated these effects graphically, as shown in Figure 8.2. To take maximum advantage of the 6-dB magnetic field enhancement shown, body-worn applications should feature magnetic antennas, in which the reactive part of their radiated complex power is positive (i.e., inductive).

Using an alternative approach, Fujimoto et al. [30] have shown that an electrically small loop antenna (effectively a magnetic dipole) has a gain advantage over other types when close to the body. Figure 8.3 shows a simple model for comparing a small loop and a short dipole at a distance *a* from the body surface. The body is treated as a solid, plane reflector, and the current in the dipole is in the opposite direction to its image. This means that the field produced by the image dipole must be subtracted from the actual dipole fields. For the loop, however, the image currents are flowing in the same direction as in the real antenna, so the image field must be added to the loop's field. Note that in Figure 8.3, the loop is shown oriented normal to the body surface for magnetic field enhancement. Analysis shows that a loop parallel to the body surface generates an image with opposing current, reducing the resultant total field. The loop in Figure 8.3 is shown as the combination of four circulating electric field vectors, equivalent to a magnetic dipole oriented along the loop axis (parallel to the body surface). According to image theory [31], the reflection of a parallel magnetic source in an electric conductor is in phase with the source, agreeing with the analysis.

Using the plane conductor model shown in Figure 8.3, Fujimoto et al. [30] derived expressions for the directional gain of a short dipole and a loop (magnetic dipole) at a distance *a* from the body surface. For a short dipole, the body-induced gain is:

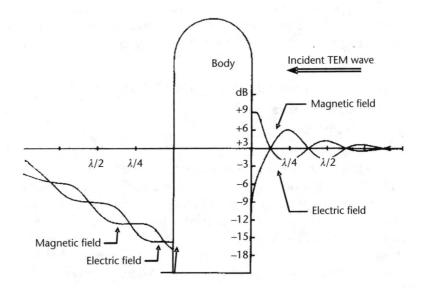

Figure 8.2   Effect of the human body on an electromagnetic field, with values given for 150 MHz. (*After:* [29].)

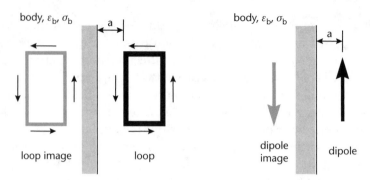

Figure 8.3   Plane conductor model for chest-mounted antennas.

$$G'_{body} = \sqrt{\frac{R_r + R_{loss}}{R_r + R_{loss} + R_m}} \cdot 2\sin\left(\frac{2\pi}{\lambda}a\right) \tag{8.6}$$

where R_r is the radiation resistance, R_{loss} is the loss resistance, and R_m is the mutual resistance between the antenna and its image. For an electrically small antenna, $R_{loss} >> R_r$, and for increasing values of a, $R_{loss} >> R_m$, reducing (8.6) to:

$$G'_{body} = 2\sin\left(\frac{2\pi}{\lambda}a\right) \tag{8.7}$$

However, for a small loop with the image in phase, the directional, body-induced gain is:

$$G_{body} = 2\cos\left(\frac{2\pi}{\lambda}a\right) \tag{8.8}$$

Equations (8.7) and (8.8) are plotted for $\lambda = 71.8$ cm (418 MHz) in Figure 8.4. For the typically small values of a found in body-worn applications, the loop (magnetic) antenna offers a significant advantage over a dipole (electric) antenna.

8.2.1   Antenna-Body Effects (Less Than 1 GHz)

8.2.1.1   Loop Antenna at 418 MHz

As described earlier, a small loop antenna is a reasonable choice for compact body-mounted telemedicine equipment. Figure 8.5 compares the FDTD-calculated azimuthal (ϕ-cut) radiation patterns of a 300-mm^2 printed loop antenna in isolation and when body-worn. A tissue-layered (skin, fat, and muscle) FDTD model was constructed to represent a 6-year-old female with a height of 1.26m and a body mass of 21.6 kg. The antenna was part of a body-worn pediatric EKG telemeter. The model consisted of 1,403,138 (94 × 59 × 253) 5-mm cubical voxels. Figure 8.5(a) shows the loop in isolation. Note that the pattern is normal to the plane of the loop (i.e., with the loop coming out of the page). When placed in the center and normal to the chest of the phantom, the pattern shown in Figure 8.5(b) developed a directional gain of 2.3 dB, and distinctive nulls in the reverse direction. The plots were normal-

Body-induced gain (dB)

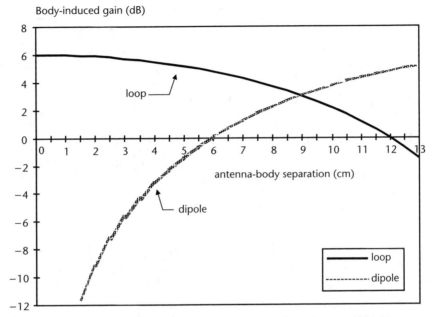

Figure 8.4   Theoretical body-induced gain for electrically small antennas at 418 MHz.

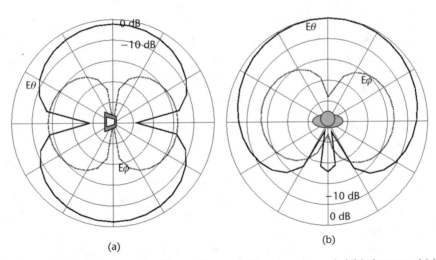

Figure 8.5   418-MHz chest-mounted loop antenna, with a three-layered child phantom: (a) loop in isolation, and (b) when body-worn (0-cm separation). (*From:* [27]. © 1999 IOP Publishing Ltd. Reprinted with permission.)

ized to the maximum field-strength result (i.e., chest-worn loop in the forward direction), and the loop was touching the chest (i.e., antenna-body separation was 0 cm).

Practical radiation pattern measurements were then recorded on an elevated test range. The measured azimuthal radiation patterns shown in Figure 8.6 were obtained during a single session, and with the same 28-kg, 1.25-m-tall female subject. Both horizontal and vertical field strengths were measured simultaneously using a crossed-element Yagi antenna. The patterns were in reasonable agreement

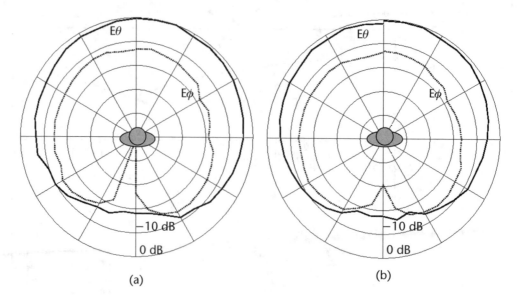

Figure 8.6 Measured azimuthal patterns for 418-MHz chest-mounted loop: (a) test #1, and (b) test #2. (*From:* [27]. © 1999 IOP Publishing Ltd. Reprinted with permission.)

with the calculated pattern in Figure 8.5(a). There was a 20-dB reduction in $E\theta$ field strength in the reverse direction, and the $E\phi$ pattern was 15 dB below $E\theta$. The patterns were measured in 10° increments, and in both cases, the discrete angle increments meant that only one of the $E\phi$ nulls was detected.

Figure 8.7 shows azimuthal power patterns for 1-, 2-, 3-, and 4-cm antenna-body separations. The patterns have been normalized to the 0-cm condition shown in Figure 8.5(b). As the separation distance increases, the forward pattern gain advantage reduces; in the reverse direction, the mean gain rises, but the nulls are still present. Figure 8.8 compares the maximum body-induced pattern gain for these results with the theory derived by Fujimoto et al. [see (8.8)]. The pattern averaged gain (PAG) for the total field ($E\theta + E\phi$) is also shown for comparison. The numerical results obtained follow the trend of, and are at most 30% (2 dB) less than, those derived analytically.

A useful metric for assessing body-worn antenna performance is the ratio of total radiated power when body-worn to total radiated power when in free-space isolation. For convenience, this is termed body-worn efficiency, η, and it represents the overall power losses in the user's body. Comparing the total power radiated by an isolated antenna with the body-worn case, the FDTD model of the 418-MHz loop reported a body-worn efficiency of 60.7% (2.2-dB body loss). Since pattern measurements were made only for the azimuthal (i.e., subject standing upright) case, the measured body-worn efficiency was estimated as 49% (3.1-dB body loss), by extrapolation from pattern averaged gain.

The antenna-body separation distance also affects radiation efficiency. Figure 8.9 illustrates that for a 418-MHz loop, efficiency increases to almost 66% as the spacing reduces, but falls to 60% when touching. This directly correlates to the variable-spacing pattern results shown in Figure 8.8, where the maximum gain of 4.4 dB occurs at 1 cm. This means that magnetic antennas should be mounted with a small

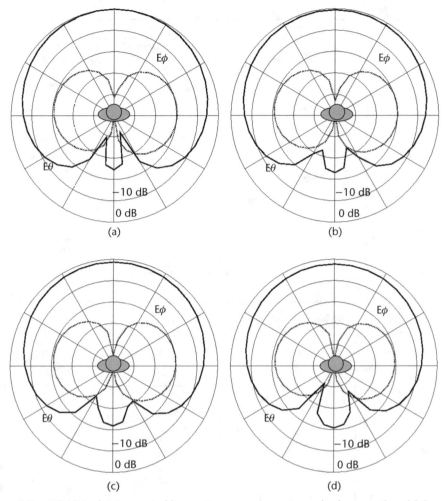

Figure 8.7   418-MHz chest-mounted loop antenna varying antenna-body separation: (a) 1 cm, (b) 2 cm, (c) 3 cm, and (d) 4 cm.

separation from the body surface, where practical, while being careful to ensure that the distance is fixed to avoid fluctuations in antenna input impedance.

8.2.1.2   Loop Antenna at 916 MHz

The calculations and measurements were repeated at 916.5 MHz. Compared with 418 MHz, a compact loop antenna is fundamentally more efficient at this higher frequency, since it is electrically larger. However, wave attenuation increases with frequency, and this increase is compounded, since the effective conductivity of body tissues also increases. Figure 8.10 compares the measured and calculated azimuthal radiation patterns; body-induced losses were greater and pattern fragmentation was more pronounced. In this case, the calculated body-worn efficiency was 10.2% (9.9 dB body loss), while the measurements reported an efficiency of 9.1% (10.4 dB body loss). These results are consistent with the higher losses in biological tissues at 916.5 MHz.

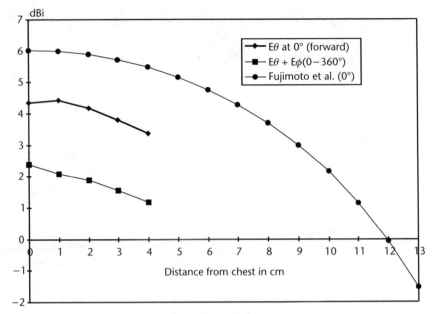

Figure 8.8   Body-induced pattern gain for a 418-MHz loop.

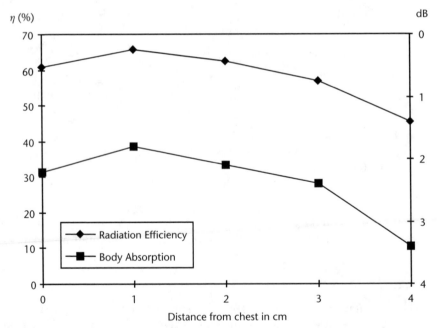

Figure 8.9   Performance variation with antenna-body spacing, with a 418-MHz loop.

8.2.2 Antenna-Body Effects (Greater Than 1 GHz)

In the microwave region and above, the increased energy absorption in biological tissues suggests that the human body will have even greater influence on the performance of nearby antennas. In Section 8.2.1, it was confirmed through modeling and measurement that the efficiency of body-worn antennas dramatically reduces with increasing frequency, resulting in a 7-dB increase in body losses at 916.5 MHz com-

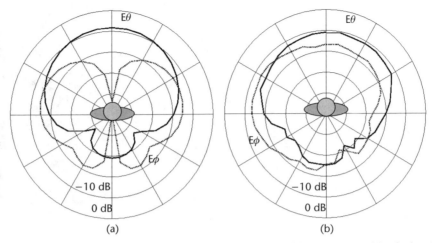

Figure 8.10 Radiation patterns for a 916.5-MHz chest-mounted loop antenna: (a) calculated, and (b) measured. (*From:* [32]. © 2001 IEE. Reprinted with permission.)

pared to 418 MHz. However, despite much higher tissue losses at 2.45 GHz, antenna performance can be surprisingly good in this band, particularly when the antenna-body spacing is $\lambda/8$ or more. Figure 8.11(a) compares measured and calculated azimuthal radiation patterns for a chest-mounted $\lambda/4$ monopole antenna mounted on a $100 \times 50 \times 25$ mm conducting box. This device was based on a biomedical transponder for in-ward patient monitoring [33]. The pattern is given in decibels relative to the calculated peak copolar gain of the antenna-body system (2.6 dBi). The measured results were recorded at 10° intervals on an outdoor elevated range.

The whole-body FDTD model in Figure 8.11(b) illustrates how, due to the slope of the chest, the feedpoint of the monopole antenna was 5 cm from the surface of the body. At 2.45 GHz, this equates to a spacing of almost $\lambda/2$, and was sufficient to substantially reduce tissue losses, leading to a calculated antenna efficiency of 49.0% (3.1-dB body loss). Extrapolating pattern averaged gain, the measured body-worn efficiency was 51% (2.9-dB body loss). Note that the phantom used for the FDTD calculations consisted of 6 distinct tissues with a voxel resolution of 5 mm, and represented a 39-year-old male, 1.8m in height, and with a body mass of 109 kg.

For practical telemedicine applications at this frequency, it is more likely that integrated antennas will be required for credit card–sized devices. For example, a 2.45 GHz inverted-F antenna with coplanar printed circuit board ground plane is shown in Figure 8.12. The F had a total length of 25 mm, with a height of 5 mm from a 40×70 mm ground plane. The feedpoint was 5 mm away from the grounding point. Figure 8.12 shows the FDTD-calculated azimuthal and vertical radiation patterns (in decibels relative to isotropic antenna) for the antenna when mounted *normal* to the chest wall of the whole-body phantom described above. The active antenna element is only 15 mm from the body surface, and the calculated antenna efficiency in this case was 25.0% (6-dB loss). The vertical radiation patterns show that the performance in the anterior direction is reasonably omnidirectional; however, in the posterior direction, gain is severely reduced, on average by more than 15

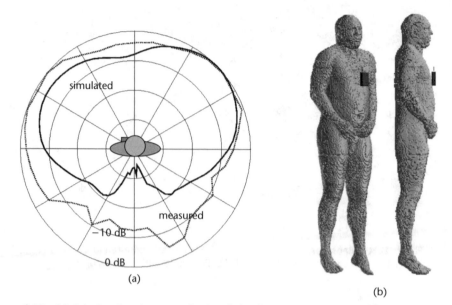

Figure 8.11 (a) Calculated and measured azimuthal radiation patterns for a chest-worn $\lambda/4$ monopole antenna on a conducting box at 2.45 GHz. (b) A 5-mm voxel whole-body representation of an adult male used for FDTD simulations. (*From:* [32]. © 2001 IEE. Reprinted with permission.)

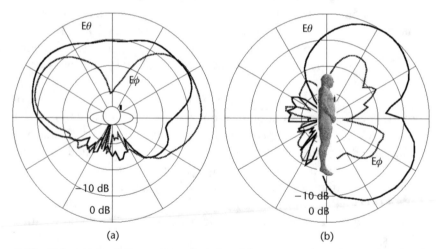

Figure 8.12 Calculated radiation patterns for printed-F antenna, oriented normal to the chest wall: (a) azimuthal, and (b) vertical. (*From:* [32]. © 2001 IEE. Reprinted with permission.)

dB. In practice, a telemedicine device is more likely to be oriented parallel with the body surface itself (e.g., in the user's pocket). The FDTD calculations were repeated for the inverted F antenna and ground plane, but in this case, the entire card was oriented parallel to the body surface. The arrangement was such that the ground plane was a minimum of 5 mm from the body, while in the region of the F antenna, there was a 10-mm separation. The radiation patterns for this configuration are presented in Figure 8.13. The azimuthal plot is strongly asymmetric, despite the antenna feedpoint being offset to the right of the user's chest in both cases. The vertical pat-

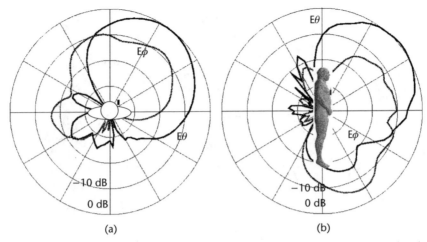

Figure 8.13 Calculated radiation patterns for printed-F antenna, oriented parallel to the chest wall: (a) azimuthal, and (b) vertical. (*From:* [32]. © 2001 IEE. Reprinted with permission.)

terns confirm the reduced gain in the posterior direction. Table 8.2 compares the merits of the two configurations.

The higher frequency WLAN technology may also be used for telemedicine. Figure 8.14 shows the FDTD calculated azimuthal radiation pattern of a hip-worn telemedicine unit operating at 5.2 GHz (vertical polarization, $E\theta$). The model included an anatomically realistic human body phantom, a conducting box (representing the patient unit), and a thin-wire dipole antenna. The overall FDTD grid was $499 \times 93 \times 154$ mm with 3.6 mm^3 voxels. The body phantom was for a 1.75-m tall adult male, and incorporated 21 tissue types. The sleeve-dipole antenna used in measurements was modeled as a center-fed 25.2-mm (0.36-mm radius) thin-wire element, and was positioned with a minimum antenna-body spacing of 14.4 mm. The high degree of separation ($>2\lambda$) reduced the overall body losses, with a corresponding FDTD-computed radiation efficiency of 83.3% at 5.2 GHz. The computed pattern was strongly directional, with a peak gain of +6.0 dBi and a through-body null of −37.9 dBi in the azimuthal plane.

8.2.2.1 Cellular

The body-shadowing effect for a cellular telephone held close to the user's head is well known and validated [34, 35]. However, in a telemedicine application, the handset must operate effectively when body-worn at waist height. This is a convenient location for fixed use over a period from 24 to 72 hours (a likely monitoring period), and allows the routing of cables to localized biomedical signal acquisition

Table 8.2 Comparison of Body-Worn Performance Characteristics for 2.45-GHz Printed-F Antenna

Configuration	Input Impedance	Body-Worn Efficiency	Peak Gain (dBi)	Range (dB)
Normal to chest wall	$82.3 - j37.2\Omega$	24.9% (6.0 dB)	+0.0	50.4
Parallel to chest wall	$73.6 - j33.0\Omega$	17.8% (7.5 dB)	+0.7	48.1

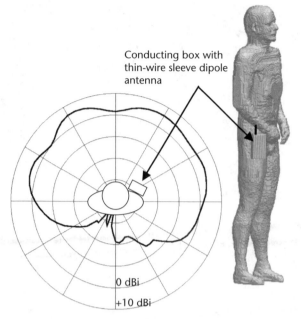

Figure 8.14 Azimuthal radiation pattern for hip-worn 5.2-GHz telemedicine unit. (*From:* [36]. © 2004 IEEE. Reprinted with permission.)

modules, as shown in Figure 8.15. The FDTD-calculated azimuthal copolar and cross-polar patterns (*Eθ* and *Eφ*, respectively) for an 1,800-MHz quarter-wavelength monopole, mounted on a conducting enclosure and placed at waist height on an adult male, are also shown in Figure 8.15. Note the through-body nulls. Close

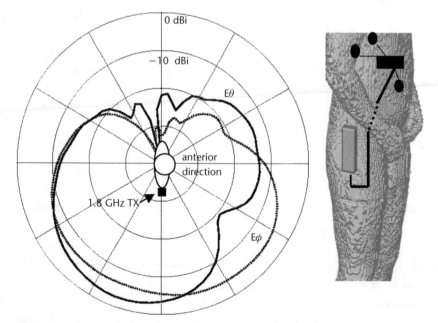

Figure 8.15 Calculated radiation patterns for an 1,800-MHz cell phone at waist height. (*From:* [37]. © 2001 IEE. Reprinted with permission.)

proximity of sensor wires to the handset's antenna is a potential cause for further service degradation and RFI to biomedical sensors.

8.2.3   Emerging Antennas

Wearable antennas for telemedicine are usually designed for maximum efficiency and omnidirectional coverage, but with WBAN schemes, minimal off-body radiation and maximum coupling between body-worn devices is required. Regardless of whether the antenna is designed for a far-field radiation mode or for a body-surface coupling mode, a major design challenge for a low-profile antenna (usually a patch structure) is a reduction of the energy lost in the body due to ground plane currents. In addition, more advanced applications may utilize individual elements in a smart antenna formulation to provide functions such as beam-steering, null-steering, frequency-selective responses, or channel capacity enhancement through MIMO.

Research interest in body-mounted antennas has rapidly increased in recent years, although the potential benefits and fundamental challenges were well understood by King over 30 years ago [38]. Since then, low-profile body-mounted antenna research has kept pace with wider developments, but there have not been any distinctive breakthroughs in the design of the wearable antenna. For example, for the shoulder-mounted PIFA structure presented by Ogawa et al. [39], the extent of the ground plane had the most significant effect on antenna performance. Furthermore, there is an increased interest in using multiple antenna elements (such as in the body-worn harness with tapered slot antennas used by Kohls et al. [40]) to provide either diversity or improved omnidirectional coverage. At first glance, microstrip patch antennas seem ideal for body-worn applications. For example, FDTD simulations have shown that for a 2.45-GHz microstrip patch of dimensions 23.8×37.4 mm (TMM6 substrate, $\varepsilon_r = 6.0$) placed on a three-layer tissue model (skin, fat, and muscle, with thicknesses of 0.7, 2.9, and 6.5 mm, respectively), there was no effect on patch resonant frequency or bandwidth, and only a small (<4%) reduction in radiation efficiency. The tissue model had a surface area of twice the patch ground plane. However, earlier research work on wearable low-profile patch antennas concluded that omnidirectional pattern coverage was not possible with one element [41]. Investigation of the near-field distribution in the TMM6-based patch design showed the expected propagating wave excitation at the radiating edges. There was also a propagating wave generated in the substrate that coupled into the biological tissues at the edge of the substrate/ground plane. In conventional patch applications, this surface wave excitation is considered undesirable, and becomes more significant as the substrate permittivity increases. However, provided that substrate losses can be minimized, the surface propagating mode may provide the basis for either better omnidirectional radiation patterns or improved performance in applications where the antenna is coupled to the body. Figure 8.16 [42] illustrates a 2.45-GHz propagating wave close to the body surface that is diffracted around the body without excessive loss of energy due to tissue coupling. The source was an inverted-F antenna, normal to the user's chest, and the corresponding radiation pattern was given in Figure 8.12.

As described by Salonen et al. [43], the use of textile antennas has also been investigated, with even cellular (GSM) examples appearing in the literature [44].

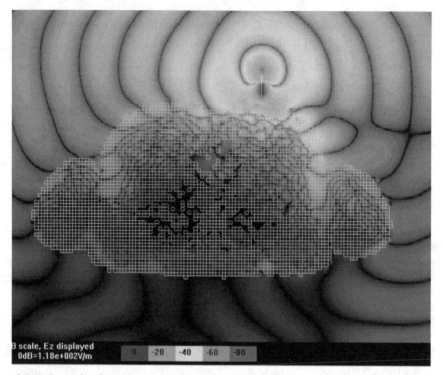

Figure 8.16   Example of propagating field pattern around the human torso at 2.45 GHz. The antenna element was a perpendicular PIFA.

However, a clothing-based textile antenna can suffer from a large variability in body separation distance as the user moves. This causes significant antenna input impedance variability that is very difficult to accommodate through matching, and may lead to rapidly changing radiation pattern fragmentation. Despite some of the claims made in the literature (e.g., impedance and radiation characteristics remain nearly unaffected, based on simple FDTD models without consideration of antenna bending, patch compression, or user movement [45]), these radiators generally suffer from extremely poor efficiency and are prone to manufacturing-related performance variations. Other body-mounted antenna research includes investigations of on-body channels for UWB applications. Alomainy et al. [46] note that an over-the-body surface propagation mode was present for a printed horn antenna mounted parallel to the body, while in Welsh et al. [47], measurements indicated a definite null in the normalized UWB antenna pattern when the latter was body-worn in large open spaces with a limited multipath.

Another important area of antenna research is the broad area of electromagnetic metamaterials. Recent advances in EBG materials, as reported by Salonen et al. [48], suggest that both the bandwidth and size constraints of wearable low-profile antennas can be lessened by incorporating advanced electromagnetic materials into the antenna element design. Of particular importance is the minimization of ground-plane currents, which cause high tissue losses in wearable patch antennas, while retaining the ability to provide the required near-field or far-field distributions for application-specific requirements. For example, Baccarelli et al. [49] derived the fundamental modal properties of surface waves on grounded metamaterial slabs, and showed that for a double negative metamaterial with $\varepsilon_r\mu_r < 1$, surface wave suppression can be obtained if the substrate height is sufficiently large. Therefore, it is very likely that innovative low-profile wearable antenna solutions based on electromagnetic metamaterials will emerge.

8.3   System Design Issues

8.3.1   Channel Effects

Radio wave propagation in indoor environments is dominated by multipath fading effects, partition shadowing, and diffraction [50]. However, both the overall path loss and fading effects tend to be greater indoors than in mobile (outdoor) environments. In addition, human body shadowing effects can be significant even within relatively small areas, such as hospital wards. For example, a body-worn 5.2-GHz measurement receiver was used to characterize the channel conditions within an empty 7m × 6m room. The receiver was worn on the hip, and had a sleeve dipole antenna. A +10-dBm CW transmitter was located 0.3m above the midpoint of the longer wall. The received power was sampled at 10 ms intervals as the test subject walked away from the transmitter (NLOS conditions). Figure 8.17 shows the measured and predicted received power profile plotted against transmitter-receiver separation. The predicted profile was generated from the FDTD model of a hip-worn terminal, as discussed in Section 8.2.1, and geometrical optics-based ray tracing. Similar measurements were obtained for parallel trajectories at 0.5-m intervals across the entire room width, providing a total of 14 NLOS and 14 LOS profiles. Averaging all of the profiles taken within the room, the measured LOS received power was 3.8 dB higher than for the measured NLOS received power. Although it does not account for fast-fading effects, this is a good estimate of the body shadowing effect that must be included in the link budget for similar installations. However, when considering individual trajectories, it was found that the body shadowing effect varied significantly, and was actually negative (i.e., NLOS power was greater than LOS power) for the trajectories at the room edges.

In terms of fading distribution, propagation theory would suggest that the channel should be Rician for the NLOS case and Rayleigh for LOS. However, the chan-

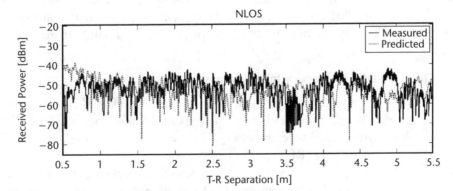

Figure 8.17   Measured and predicted multipath fading in a 42-m^2 room for a moving body-worn receiver at 5.2 GHz.

nel conditions were found to be log-normal for both LOS and NLOS conditions (Figure 8.18), despite the short distances involved in the trajectories. Note that the cumulative distribution functions in Figure 8.18 were calculated from received power levels with respect to the local mean. The simulations based on the FDTD ray tracing model were in good agreement with the LOS results, but were less effective in modeling NLOS conditions.

Wideband channel characteristics are also an important aspect of system design, and over even relatively small distances, these characteristics can change rapidly. Figure 8.19 shows the normalized PDP obtained over a 1-m measurement distance in the main service corridor of Coleraine General Hospital, Northern Ireland, United Kingdom. The PDP was obtained using a 2,340-MHz pulse transmitter (7.7 ns pulses with a peak power of 2W and a pulse repetition frequency of 150 Hz). Both transmit and receive antennas were omnidirectional discones. The multipath nature of the channel is evident, with significant time dispersion as echoes that were received as much as 300 ns after the main pulse.

A further consideration is the temporal fading caused by the movement of people and equipment in the local environment. Figure 8.20(a) shows the floor plan of the surgical ward at Coleraine General Hospital, United Kingdom. Using a stationary 2,340-MHz CW transmitter (TX) and receiver (RX), the path loss was recorded at 10-ms intervals for 135 minutes. Figure 8.20(b) shows that for this 45-m link, fades of up to 40 dB were common, with a noticeable reduction in fading rate during periods of less activity, particularly at morning break and lunchtime.

Channel propagation is also of concern for short-range, home-based telecare and telemedicine systems. In these applications, the patient device or sensor network often has a wireless link to a single (off-body) controller, where data may be logged, processed, or relayed. Unlike commercial wireless networking installations, where access points are routinely mounted high on walls or on the ceiling, the controller (or base station) in a home typically may be placed at a low height to facilitate connection to the ac power supply, and if required, connection to the PSTN for dial-up or ADSL-based communication. The low height effectively reduces coverage and makes the link very susceptible to fading and shadowing effects. There have not been many studies of UHF propagation within home environments. However, in [51], it was suggested that a 10-dB increase in transmit power is required at 868 MHz to provide the same availability as a corresponding 433-MHz system, with an additional 20 dB required at 2.45 GHz. These results were based on measurements made in five different domestic dwellings with full-size isolated antennas, and they do not take into account body effects or electrically small antennas.

8.3.2 Radio Frequency Interference

Other electromagnetic considerations for wearable telemedicine devices include the effects of unintentional radiators, such as equipment cables [52], and user items, such as wire-framed spectacles [53]. These considerations create a risk of radio frequency interference (RFI) for biomedical sensors.

Over the past few years, there has been a worldwide growth in the number of mobile radio users, which implies a high probability of usage near susceptible medical devices. RFI problems have been reported in hospital environments for EKG

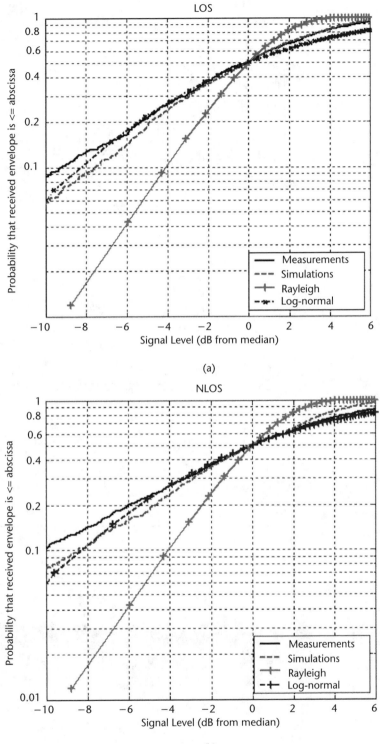

Figure 8.18 Cumulative distribution functions for simulations and measurements of received power relative to the local mean in a 42-m² room, with a 5.2-GHz body-worn receiver: (a) LOS, and (b) NLOS.

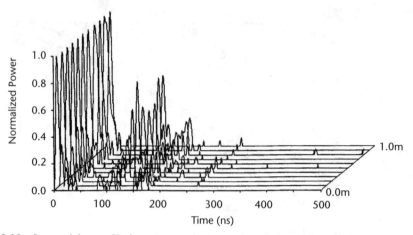

Figure 8.19   Power delay profile for a 1-m path along a hospital corridor at 2,340 MHz.

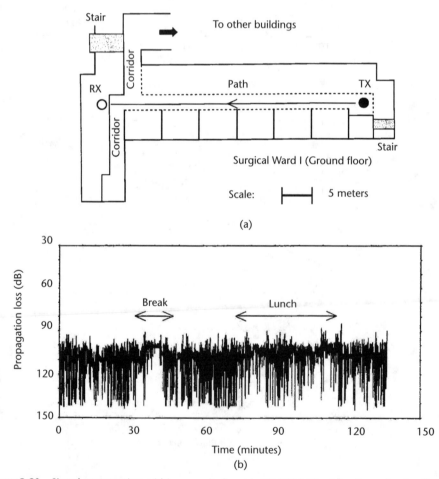

(a)

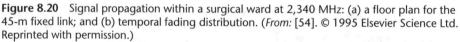

(b)

Figure 8.20   Signal propagation within a surgical ward at 2,340 MHz: (a) a floor plan for the 45-m fixed link; and (b) temporal fading distribution. (*From:* [54]. © 1995 Elsevier Science Ltd. Reprinted with permission.)

monitors and infusion pumps, caused by GSM cell phones placed in their near vicinity (≤20 cm) [11]. In some cases, significant interference to the EKG output wave was observed, causing its information content to be irrecoverable. Similar problems can also arise in radio-based personal telemedicine systems. For example, emergency-service workers can wear appropriate instrumentation interfaced to their medium-power personal radio (PR), or a separate and dedicated body-worn patient monitor.

Figure 8.21 shows one possible RFI scenario, consisting of a chest-worn EKG telemeter and a nearby λ/2 dipole antenna representing a 435-MHz PR [55]. In an FDTD analysis, the body torso was represented by a homogeneous muscle slab measuring 70 × 30 × 15 cm. The EKG monitor wire was 10 mm from the slab surface, and its length was 188 mm, representative of that used in practice. The peak current induced in the monitor wire was investigated by varying the horizontal separation between the dipole and wire, from 10 to 100 mm. The maximum induced current, in a vertical monitor wire at a separation of 2 cm, normalized to a dipole feedpoint drive current of 100 mA, was 22 mA, as shown in Figure 8.22. The current then decays with separation as the feed resistance grows. To avoid staircasing errors, the FDTD model of the tilted monitor wire used special Dey-Mittra update equations, giving a maximum induced RF current of 7 mA. In both instances, the coupled current magnitudes were well within the RFI detection threshold of the EKG circuitry, whether by unintentional envelope demodulation or by amplifier saturation.

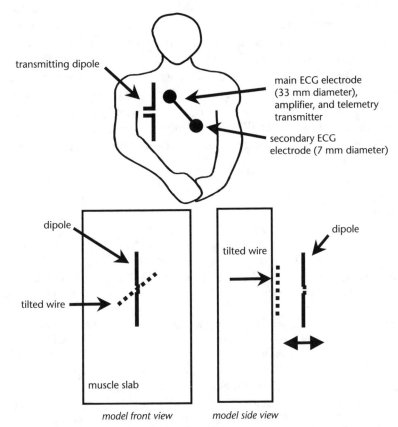

Figure 8.21 RFI example with computational model.

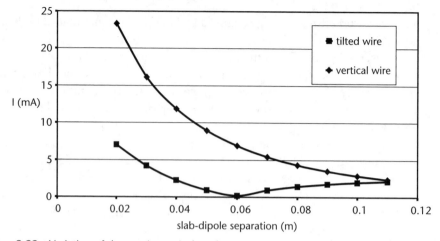

Figure 8.22 Variation of the maximum induced monitor wire current versus separation distance, for a dipole feed of 100 mA. (*From:* [55]. © 2005 IEE. Reprinted with permission.)

8.4 Conclusion

Antennas and propagation are both important aspects of wearable telemedicine system design. In particular, the choice of operating frequency will be influenced by constraints in the antenna size, and by application-specific factors such as bandwidth and error performance requirements. Likewise, while there are regulatory limitations and dc power considerations concerning transmitter power, the RF link budget will be dominated by estimates of propagation effects, such as fading and body shadowing. Antenna requirements for patient-worn devices is very much a matter of reducing body losses and ensuring that there is a sufficient proportion of the transmitted power in the desired propagating mode, which in some body area networks may be directed over the body surface rather than away from the user's body. Ultimately, on-body telemedicine remains an important application for body-centric communications.

References

[1]   Parsons, D. F., "The Impact of Information Technology in Health Care: A Practitioner's Perspective," *Telematics and Inform.*, Vol. 11, 1994, pp. 127–135.

[2]   Park, J., et al., "4-Subject 4-Channel Optical Telemetry System for Use in Electrocardiograms," *IEEE Int. Conference on Electronics, Circuits and Systems*, Vol. 3, 1998, pp. 251–254.

[3]   Wei, K., "Design and Initial Evaluation of an Implantable Sonomicrometer and CW Doppler Flowmeter for Simultaneous Recordings with a Multichannel Telemetry System," *IEEE Trans. on Biomedical Engineering*, Vol. 52, No. 7, July 2005, pp. 1365–1367.

[4]   Pattichis, C. S., "Wireless Telemedicine Systems: An Overview," *IEEE Antennas and Propagation Magazine*, Vol. 44, No. 2, April 2002, pp. 143–153.

[5]   Barlow, J., S. Bayer, and R. Curry, "The Design of Pilot Telecare Projects and Their Integration into Mainstream Service Delivery," *J. of Telemedicine and Telecare*, Vol. 9, Suppl. 1, 2003, pp. S1–S3.

[6]   ANSI/IEEE Std 802.11, 1999 Edition, Part 11: Wireless LAN Medium Access Control (MAC) and Physical Layer (PHY) Specifications, IEEE, 1999.

[7]   Scanlon, W. G., et al., "Low-Power Radio Telemetry: The Potential for Remote Patient Monitoring," *J. of Telemedicine and Telecare*, Vol. 2, No. 4, December 1996, pp. 185–191.

[8]   Anliker, U., et al., "AMON: A Wearable Multiparameter Medical Monitoring and Alert System," *IEEE Trans. on Information Technology in Biomedicine*, Vol. 8, No. 4, December 2004, pp. 415–427.

[9]   Einthoven, W., "Le Telecardiogramme," *Arch. Int. de Physiol.*, Vol. 4, 1906, pp. 132–164. Translated into English, *Am. Heart J.*, Vol. 53, 1957, pp. 602–615.

[10] Tompkins, W. J., (ed.), *Biomedical Digital Signal Processing*, Upper Saddle River, NJ: Prentice Hall, 1993.

[11] Morrissey, J. J., M. Swicord, and Q. Balzano, "Characterization of Electromagnetic Interference of Medical Devices in the Hospital Due to Cell Phones," *Health Physics*, Vol. 82, No. 1, January 2002, pp. 45–51.

[12] http://wireless.fcc.gov/services/personal/medtelemetry/.

[13] Lin, Y. -H., et al., "A Wireless PDA-Based Physiological Monitoring System for Patient Transport," *IEEE Trans. on Information Technology in Biomedicine*, Vol. 8, No. 4, December 2004, pp. 439–447.

[14] Crumley, G. C., et al., "The Design and Performance of a 2.5-GHz Telecommand Link for Wireless Biomedical Monitoring," *IEEE Trans. on Information Technology in Biomedicine*, Vol. 4, No. 4, December 2000, pp. 285–291.

[15] Rasid, M. F. A., and B. Woodward, "Bluetooth Telemedicine Processor for Multichannel Biomedical Signal Transmission Via Mobile Cellular Networks," *IEEE Trans. on Information Technology in Biomedicine*, Vol. 9, No. 1, March 2005, pp. 35–43.

[16] Federal Communications Commission, RM-11271, Petition for Rulemaking, "Amendment of Parts 2 and 95 of the Commission's Rules to Establish the Medical Data Service at 401–402 MHz and 405–406 MHz," Report No. 2725, August 24, 2005.

[17] Anliker, U., et al., "A Systematic Approach to the Design of Distributed Wearable Systems," *IEEE Trans. on Computers*, Vol. 53, No. 8, August 2004, pp. 1017–1033.

[18] Deasy, T. P., and W. G. Scanlon, "Stepwise Algorithms for Improving the Accuracy of Both Deterministic and Probabilistic Methods in WLAN-Based Indoor User Localisation," *Intl. J. of Wireless Information Networks*, Vol. 11, No. 4, October 2004, pp. 207–216.

[19] Ng, J. W. P., "Ubiquitous Healthcare Localisation Schemes," *Proc. 7th Int. Workshop Enterprise Networking and Computing in Healthcare Industry (HEALTHCOM)*, June 2005, pp. 156–161.

[20] Liszka, K. J., et al., "Keeping a Beat on the Heart," *IEEE Pervasive Computing*, Vol. 3, No. 4, October–December 2004, pp. 42–49.

[21] Kimmich, H. P., "Biotelemetry," in *Encyclopaedia of Medical Devices and Instrumentation*, Vol. 1, J. G. Webster, (ed.), New York: John Wiley & Sons, 1988, pp. 409–425.

[22] King, R. W. P., S. Prasad, and B. H. Sandler, "Transponder Antennas in and Near a Three-Layered Body," *IEEE Trans. on Microwave Theory and Techniques*, Vol. 28, 1980, pp. 586–596.

[23] Neukomm, P. A., "Small Body-Mounted Antennas: The Influence of the Human Body on the Azimuthal Radiation Pattern in the Frequency Range 50 to 1000 MHz," *Biotelemetry IV, Proc. 4th Int. Symp. Biotelemetry*, H. -J. Klewe and H. P. Kimmich, (eds.), 1978, pp. 41–44.

[24] Amlaner, C. J., "The Design of Antennas for Use in Radio Telemetry," in *A Handbook on Biotelemetry and Radio Tracking*, C. J. Amlaner and D. W. Macdonald, (eds.), Oxford, England: Pergamon Press, 1980.

[25] Scanlon, W. G., N. E. Evans, and M. Rollins, "Antenna-Body Interaction Effects in a 418 MHz Radio Telemeter for Infant Use," *18th IEEE Engineering in Medicine and Biology Soc. Conference*, Vol. 1, Amsterdam, 1996, pp. 278–279.

[26] Scanlon, W. G., and N. E. Evans, "Body-Surface Mounted Antenna Modelling for Biotelemetry Using FDTD with Homogeneous, Two- and Three-Layer Phantoms," *10th Int. Conference on Antennas and Propagation*, IEE Conf. Pub. 436, Vol. 1, 1997, pp. 342–345.

[27] Scanlon, W. G., N. E. Evans, and J. B. Burns, "FDTD Analysis of Close-Coupled 418 MHz Radiating Devices for Human Biotelemetry," *Physics in Medicine and Biology*, Vol. 44, No. 2, February 1999, pp. 335–345.

[28] Burberry, R. A., "Bodyborne Antennas," Ch. 10 in *VHF and UHF Antennas*, London: Peter Perigrinus Ltd., 1992, pp. 174–186.

[29] Mumford, R., Q. Balzano, and T. Taga, "Land Mobile Antenna Systems II: Pagers, Portable Phones and Safety," Ch. 4 in *Mobile Antenna Systems Handbook*, K. Fujimoto and J. R. James, (eds.), Norwood, MA: Artech House, 1994.

[30] Fujimoto, K., et al., *Small Antennas*, Hertz, U.K.: Research Studies Press, 1987.

[31] Balanis, C. A., *Advanced Engineering Electromagnetics*, New York: John Wiley & Sons, 1989, p. 317.

[32] Scanlon, W. G., and N. E. Evans, "Numerical Analysis of Bodyworn UHF Antenna Systems," *IEE Electronics and Communications Engineering Journal*, Vol. 13, No. 2, April 2001, pp. 53–64.

[33] Crumley, G. C., and N. E. Evans, "An Experimental RF Transponder for Biomedical Signal Acquisition," *IEE Coll. on RF and Microwave Circuits for Commercial Wireless Applications*, Savoy Place, London, 1997, pp. 10/1–10/5.

[34] Toftgard, J., S. N. Hornselth, and J. B. Andersen, "Effects on Portable Antennas of the Presence of a Person," *IEEE Trans. on Antennas and Propagation*, Vol. 41, No. 6, June 1993, pp. 739–746.

[35] Balzano, Q., O. Garay, and T. J. Manning, "Electromagnetic Energy Exposure of Simulated Users of Portable Cellular Telephones," *IEEE Trans. on Vehicular Technology*, Vol. 44, No. 3, August 1995, pp. 390–403.

[36] Ziri-Castro, K. I., W. G. Scanlon, and N. E. Evans, "Indoor Radio Channel Characterisation and Modelling for a 5.2-GHz Bodyworn Receiver," *IEEE Antennas and Wireless Propagation Letters*, Vol. 3, No. 11, 2004, pp. 219–222.

[37] Troulis, S. E., N. E. Evans, and W. G. Scanlon, "Propagation Issues Affecting the Deployment of GSM 1800-Based Personal Telemedicine Equipment," *11th Int. Conference on Antennas and Propagation*, (IEE Conf. Publ. No. 480), Vol. 1, Manchester, 2001, pp. 142–145.

[38] King, H. E., "Characteristics of body-Mounted Antennas for Personal Radio Sets," *IEEE Trans. on Antennas and Propagation*, Vol. 23, March 1975, pp. 242–244.

[39] Ogawa, K., T. Uwano, and M. Takahashi, "A Shoulder-Mounted Planar Antenna for Mobile Radio Applications," *IEEE Trans. on Vehicular Technology*, Vol. 49, No. 3, May 2000, pp. 1041–1044.

[40] Kohls, E. C., et al., "A Multi-Band Body-Worn Antenna Vest," *IEEE Antennas and Propagation Society Symp.*, Vol. 1, June 2004, pp. 447–450.

[41] Jayasundere, N., and T. S. M. Maclean, "Omnidirectional Radiation Patterns from Body Mounted Microstrip Antennas," *Proc. 6th IEE Int. Conference on Antennas and Propagation*, 1989, pp. 187–190.

[42] Scanlon, W. G., "Body-Worn Antennas for ISM-Band Applications Including Bluetooth," *IEE Coll. Integrated and Miniaturized Antenna Technologies for Asset Tracking Applications*, IEE00/065, November 2000, pp. 5.1–5.5.

[43] Salonen, P., et al., "Dual-Band Wearable Textile Antenna," *IEEE Antennas and Propagation Society Symp.*, Vol. 1, June 2004, pp. 463–466.

[44] Massey, P. J., "GSM Fabric Antenna for Mobile Phones Integrated Within Clothing," *IEEE Antennas and Propagation Society Symp.*, July 2001, pp. 452–455.

[45]   Salonen, P., Y. Rahmat-Samii, and M. Kivikoski, "Wearable Antennas in the Vicinity of Human Body," *IEEE Antennas and Propagation Society Symp.*, Vol. 1, June 2004, pp. 467–470.

[46]   Alomainy, Y., et al., "Comparison Between Two Different Antennas for UWB On-Body Propagation Measurements," *IEEE Antenna Wireless Prop. Letters*, Vol. 4, 2005, pp. 31–34.

[47]   Welsh, T. B., et al., "The Effects of the Human Body on UWB Signal Propagation in an Indoor Environment," *IEEE J. of Sel. Areas Comms.*, Vol. 20, No. 9, December 2002, pp. 1778–1782.

[48]   Salonen, P., M. Keskilammi, and L. Sydanheimo, "A Low-Cost 2.45 GHz Photonic Band-Gap Patch Antenna for Wearable Systems," *11th IEE Int. Conference on Antennas and Propagation*, Vol. 2, April 2001, pp. 719–723.

[49]   Baccarelli, P., et al., "Fundamental Modal Properties of Surface Waves on Metamaterial Grounded Slabs," *IEEE Trans. on Microwave Theory and Techniques*, Vol. 53, No. 4, April 2005, pp. 1431–1442.

[50]   Hashemi, H., "The Indoor Radio Propagation Channel," *IEEE Proc.*, Vol. 81, No. 7, 1993, pp. 943–968.

[51]   van Loon, L. J. W., "Mobile In-Home UHF Radio Propagation for Short-Range Devices," *IEEE Antennas and Propagation Magazine*, Vol. 41, No. 2, April 1999, pp. 37–40.

[52]   Troulis, S. E., W. G. Scanlon, and N. E. Evans, "Effect of a Hands-Free Wire on Specific Absorption Rate for a Waist-Mounted 1.8 GHz Cellular Telephone Handset," *Physics in Medicine and Biology*, Vol. 48, No. 12, June 2003, pp. 1675–1684.

[53]   Troulis, S. E., et al., "Influence of Wire-Framed Spectacles on Specific Absorption Rate Within Human Head for 450 MHz Personal Radio Handsets," *Electronics Letters*, Vol. 39, No. 23, November 2003, pp. 1679–1680.

[54]   Wang, L. Q., "Fading Characteristics of a 2.3 GHz Radio Telemetry Channel in a Hospital Building," *Medical Engineering and Physics*, Vol. 17, No. 3, 1995, pp. 226–231.

[55]   Ball, A. D., et al., "RF Interference (RFI) Between UHF Personal Radios and Biomedical Monitoring Sensors," IEE Seminar on Telemetry and Telematics, London, April 2005, pp. 4/1 – 4/5.

CHAPTER 9

Medical Implant Communication Systems

Anders Johansson, Anders Karlsson, William Scanlon, Noel Evans, and Yahya Rahmat-Samii

9.1 Introduction

There exists a multitude of electronic implants today. Heart pacemakers and cochlea implants are two examples. With the first implantation of a self-contained heart pacemaker into a human by Åke Senning in 1958 [1], the use of pacemakers has grown to a market of over 600,000 pacemakers per year [2]. The pacemakers have been developed so that they not only are able to correct heart block and arrhythmias, but also, in some versions, are able to defibrillate the heart into a normal state [3]. Cochlea implants are another big success of implant technology [4]. The use of these implants gives deaf patients a level of hearing that allows communication by the spoken word. The cochlea implant uses an external microphone placed at the ear, and transmits the sound as an electromagnetic signal through the skin to an electrode that is connected to the cochlear nerve in the cochlea. There are other electrical implants that have been deployed, or are under development. These include implantable drug pumps [5], vagus nerve stimulators for epilepsy treatment, deep brain stimulators for Parkinson treatment [4], muscle actuators for paraplegic patients [6], and so forth.

After an electronic implant has been placed inside the body, there is a need for communication with the implant. Different operating parameters of the implant may have to be changed, and diagnostic data may be read out from the implant. The advances in memory technology also make it possible for implants to store large amounts of data to be transferred to the treating physician. The communication with the implants is typically done over an inductive link. The inductive coupling between two coils, one external and one inside the pacemaker case, is used to transfer data to and from the implant.

There are a number of advantages if the communication with the implant can be moved to a higher carrier frequency than the inductive link presently used. These include an increase in bandwidth, which makes it possible to achieve a higher bit rate; and a propagating electromagnetic wave, which makes the system usable at longer ranges. A longer communication range makes a number of new user scenarios possible. One example is that some patients may require more frequent checks than can be made practically at the hospital. A home care unit could be placed in the patient's home. The unit communicates with the medical implant, and sends regular reports to the physician at the hospital, via the telephone system or Internet. The inductive technology is not well suited for this application, since the patient must place the external coil fairly accurately and keep it there for some period of time. RF technology would instead make it possible to place the home care unit at the bedside table and to read data every night when the patient is sleeping, thus making the surveillance more convenient. One example of such a system is the Biotronic Home Monitoring System, in which the pacemaker transmits data to a small external unit that can be worn at the belt [7]. This unit is also equipped with a GSM telephone that relays the data to the physician.

Future applications also include communication between separate implants in the same patient. This could be used for controlling a pacemaker from neural signals [8] or for controlling a limb of a spinal cord–injured patient with signals from the brain. There is also a possibility of communication between implants in different patients. The practical use of this is not obvious, but there are possibilities of mesh networking between different implants across patients in order to increase range and computational capabilities.

9.1.1 Inductive Coupling

The most common electromagnetic link to medical implants is the inductive link. A pacemaker incorporates a small coil inside the closed metal housing, and an external coil is placed on the chest of the patient, on top of the implanted pacemaker, as in Figure 9.1. The inductive coupling between the colinear coils is used as the communication channel.

The inductive link uses a carrier frequency between 9 and 315 kHz, and transmits data at a speed of up to 512 Kbps [3]. The range of communication is limited to the "touch" range; that is, the external coil housing must touch the patient's chest. The placement is often guided by indicators on the external coil, since the link is sensitive to the position of the external coil. The inductive link is also used for implantable radio frequency identification (RFID) tags [9] and the BION [10] system for electrical stimulation of muscles by a small implant. The main drawback of the inductive link is that the low frequency limits the available bandwidth and

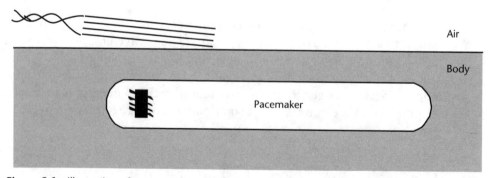

Figure 9.1 Illustration of a pacemaker with an internal telemetry coil and an external coil, which communicate by inductive coupling.

results in a low data rate. The external coil must be placed fairly accurately in order to get a reliable link, adding to the complexity of the communication procedure.

9.1.2 MICS Standard

The European Telecommunications Standards Institute (ETSI) [11] has standardized the Medical Implant Communication System (MICS) in [12]. A similar standard for the United States has been taken by the FCC [13]. The ETSI document lists two principal fields of application for the standard. The first field is for telecommunication between a base station and an implanted device. The second field is for telecommunication between medical implants within the same body. The standard does not explicitly mention a third possible use: telecommunication between medical implants in different bodies.

The frequency band allocated is from 402 to 405 MHz. The maximum emission bandwidth to be occupied is 300 kHz. The maximum bandwidth is for the complete session. If the system uses separate frequencies for uplinks and downlinks, then the two link bandwidths must not add up to more than 300 kHz. In order to get high data throughput, this implies that a half-duplex scheme should be adopted, in which only one device transmits at a time. If full duplex is necessary, the available bandwidth for transmission in each direction will be less, implying a lower data bandwidth for each direction. Note that in the case of a half-duplex solution, the uplink and the downlink do not have to share the same frequency band. Separate RX and TX bands, each with a bandwidth of 300 kHz, may be used, as long as they are not used simultaneously.

The 300-kHz bandwidth is an emission limit; the power at the band edges has to be 20 dB below the maximum level of the modulated output. The resolution bandwidth of the measurement should be 1% of the emission bandwidth of the device under test. The maximum power limit is set to 25 μW equivalent radiated power (ERP). The maximum field strength in any direction should be equal to, or lower than, what a resonant dipole would give in its maximum direction at the same distance, with the dipole being fed with a signal of 25 μW. This is to be measured with the medical implant inside a human torso simulator. There is some confusion about the power level. The ITU-R recommendation [14] sets a level of 25 μW equivalent isotropic radiated power (EIRP), which equals a level 2.15 dB lower than the ERP level set in the ETSI MICS-standard. The FCC has set the limit to EIRP = 25 μW in the United States [15], and the same level is proposed for Australia [15].

The ETSI MICS standard test procedure measures the ERP from the implant placed in a torso simulator. There is no simulator standardized for implants primarily used in arms, head, or legs. According to the standard, all implants, regardless of their final position in the body, should be tested in the same position in the same torso simulator.

The frequency band specified for MICS is already in use. The Meteorological Aids Service (METAIDS), which primarily is used by weather balloons transmitting data down to the Earth, uses the same spectrum allocation. For this reason, the MICS system is specified to be used only indoors. This is easily controllable if a fixed device that is placed indoors starts the communication sessions. If the link is used locally on the patient, between the implant and a body-worn device, or between implants in the same patient, it will be harder to fulfill this requirement.

9.1.3   2.4-GHz ISM Band

The 2.4-GHz ISM band potentially can be used for medical implant communication. It is the same band that is used today by a variety of computer equipment services (e.g., WiFi and Bluetooth). In addition, cordless telephones and household microwave ovens operate in this frequency band.

According to ETSI EN 300 328 [11], the maximum EIRP is −10 dBW (100 mW). The system should be spread spectrum, either frequency hoping spread spectrum (FHSS) or direct sequence spread spectrum (DSSS). In the case of FHSS, at least 15 separate nonoverlapping channels should be used. In the case of DSSS, the maximum power density is −20 dBW/MHz EIRP. The frequency band available is from 2.4000 to 2.4835 GHz. In the United States, the FCC limit in the same band is up to 1W, depending on the frequency hopping scheme.

The test protocol described in EN 300 328 is not intended for implanted devices. As an example, the protocol states that the batteries should be removed during testing, and the device should be run from a test power source. This is very hard to implement in a pacemaker that is welded airtight during the manufacturing process. Neither is any provision given for a human phantom of any kind.

One disadvantage with this band is that it is shared with many other users. This places great demand on interoperability and security. The attenuation in the body of the radio waves is also higher than at 400 MHz.

9.2   Lossy Dispersive Media

It is well known that an object onto which an antenna is attached influences the performance of the antenna. If the antenna is covered in order to protect it from the environment (e.g., with a radome), this will also affect its performance. When we insert an antenna into an object, such as is the case with a medical implant with an antenna inserted into a patient, we cannot separate the antenna from the surrounding object. This requirement is only loosened if the wavelength is much shorter than the size of the object; we then have to include only the parts of the object that are close to the antenna. It follows that the body covering the implanted antenna has to be accounted for when evaluating the far-field radiation characteristics of an antenna operating in the MICS band. At 403.5 MHz, the wavelength is 0.74m in air and 0.09m in the body. In a sense, the body will approximately be a very large, lossy, nonstationary radome that extends all the way from the absolute near zone of the antenna to the far zone, at least in some directions. Thus, we cannot discuss or design the antenna without investigating the electromagnetic properties of the body. For the same reason, we cannot evaluate the absolute influence of the body without discussing a certain antenna implementation.

9.2.1   Matter

In order to investigate the design of implanted antennas for higher frequencies, we need to define the electromagnetic properties of the materials. Classical antenna theory mainly deals with antennas placed in a vacuum or in air; that is, antennas that are placed in a nonconducting environment with a permittivity of $\varepsilon_0 = 8.854 \cdot 10^{-12}$

F/m. When we place the radiating structure in a material with a higher permittivity and nonzero conductivity, some of the theory must be revisited in order to revise the usual simplifications used in antenna design.

The permittivity ε and the conductivity σ are complex quantities that are expressed in their real and imaginary parts as

$$\varepsilon = \varepsilon' - j\varepsilon'' \tag{9.1}$$

$$\sigma = \sigma' - j\sigma'' \tag{9.2}$$

The complex permittivity ε_c of a medium is then defined as

$$\varepsilon_c = \varepsilon_e - j\frac{\sigma_e}{\omega} \tag{9.3}$$

Here, the effective permittivity ε_e and the effective conductivity σ_e are defined as

$$\varepsilon_e = \varepsilon' - \frac{\sigma''}{\omega} \tag{9.4}$$

$$\sigma_e = \sigma' + \omega\varepsilon'' \tag{9.5}$$

The permittivity ε_e is often scaled with the permittivity of vacuum $\varepsilon_0 = 8.854 \cdot 10^{-12}$, as in

$$\varepsilon_{er} = \frac{\varepsilon_e}{\varepsilon_0} \tag{9.6}$$

The loss due to conductivity in the matter is often expressed as a dissipation factor Diss, or a loss tangent $\tan\delta$, defined as

$$Diss = \tan\delta = -\frac{\text{Im}[\varepsilon_c]}{\text{Re}[\varepsilon_c]} = \frac{\sigma_e}{\omega\varepsilon_e} \tag{9.7}$$

where Re[ ] and Im[ ] denote real and imaginary parts, respectively.

9.2.2 Material Data and Measurements

It is essential to use accurate values of the dielectric parameters for the human body when we calculate or simulate implanted antennas. The accurate values are also necessary when we design phantoms for use in measurements of complete systems. There is much data available, mainly from the work of Gabriel [16]. This data is measured from human cadaver tissues. When we measure the permittivity of a material, we get the complex permittivity ε_c. By measuring only at a single frequency, we cannot separate the conductivity (σ/ω) from the lossy imaginary permittivity ε''. Measurement probes, such as the Agilent 89010, usually give the real part ε_r and the loss tangent $\tan\delta$.

9.2.2.1 Tissue Data

The effective permittivity ε_{er} and conductivity σ_e of different human tissues that are most relevant for subcutaneous medical implants are given in Table 9.1. The data is given for the frequencies of 403.5 MHz and 2.45 GHz, and is taken from [16]. Note that fat tissue is markedly different from both skin and muscle tissue, in that it has a much lower permittivity and conductivity.

9.2.2.2 Simulated Tissues

In order to test antenna performance of an implanted antenna in the laboratory, it is practical to use simulated instead of real tissues. The main benefit of the simulated tissues (besides the ethical issues) is that they are stable at room temperature and do not need to be kept cool in order not to deteriorate. The simplest form of simulated tissues is a liquid form. The tissues are of the same design as those used for measurement of the SAR in the evaluation of radiation from mobile handsets. The MICS standard references an article [17] in which four different materials are defined. The article gives recipes for making tissue-simulating liquids representing muscle tissue, brain tissue, and lung tissue. In addition, a recipe for making a material simulating bone suitable for casting is given. The ingredients for muscle and brain tissue simulations at 403.5 MHz are given in Table 9.2. HEC is the abbreviation for hydroxyethylcelloluse, which is an inert substance that absorbs water and increases the viscosity of a solution.

By comparing Tables 9.1 and 9.3, we see that there are differences in the values of the dielectric properties. Since the use of the tissue-simulating liquid is stipulated in the test protocols for SAR and, in the case of MICS, for EIRP measurements, the dielectric properties have to be taken into account in the design procedure, especially when simulations are used.

9.2.3 Phantoms

In order to evaluate antenna designs and make link budget calculations, we need ways to simulate the human body. There are two fundamental classes of such models, or phantoms: numerical ones used for simulations, and physical ones used for measurements.

Table 9.1 Dielectric Parameters for Human Tissue

Frequency	403.5 MHz		2.45 GHz	
Tissue	ε_{er}	σ_e (S/m)	ε_{er}	σ_e (S/m)
Muscle	57.1	0.797	52.7	1.7
Fat (noninfiltrated)	5.6	0.041	5.3	0.1
Skin (dry)	46.7	0.690	38.0	1.5
Skin (wet)	49.8	0.670	42.8	1.6
Bone cancellous	22.4	0.235	18.5	0.8
Brain gray matter	57.4	0.739	48.9	1.8
Brain white matter	42.0	0.445	36.2	1.2

Table 9.2 Recipes for Tissue-Simulating Liquids

Tissue	Water	Sugar	Salt (NaCl)	HEC
Muscle	52.4%	45.0%	1.4%	1.0%
Brain	40.4%	56.0%	2.5%	1.0%

Table 9.3 Permittivity and Conductivity at 403.5 MHz for the Simulated Tissue Materials

Material	ε_{er}	σ_e (S/m)
Muscle	62.5	0.9
Brain	50.3	0.75

The most complex phantoms use high-resolution scans of the human body as the input data. The most well-known datasets are available through the U.S. "Visible Human" project [18], in which a male has been scanned in 1-mm slices with a cross-section resolution of 0.33 mm, and a female has been scanned in 0.33-mm slices with a cross-section resolution of 0.33 mm. Similar datasets, which are generated from other human subjects, also exist. These datasets model the human body to a very high accuracy. The datasets are primarily used for numerical simulations with FDTD codes. If the datasets are directly used as the input to the numerical FDTD-simulator, they will use a lot of memory and computation time. It is therefore common to use a variable size of the discretization, with the finest resolution in the areas where the actual antenna is to be placed [19, 20]. Most of the datasets are in a standard pose with the phantoms lying down with the arms at the sides, but phantoms also exist that have moveable limbs. Two limitations of phantoms must be considered: (1) the resolution of a phantom might be too coarse to accurately model the environment in direct contact with an implant; and (2) the implant inside the actual human body displaces tissue, rather than replacing it.

A useful simplification for some applications is the use of an anthropomorphic representation of the human body as viewed from the outside, but with a homogenous inside—a mannequin doll filled with a tissue-simulation liquid. This is commonly used in the mobile phone industry, which makes use of physical head-and-shoulder phantoms when evaluating the performance of mobile phones. The homogenous phantoms are simpler to generate numerically. These phantoms are commercially available in different body shapes and ages [21]. Such tools often have the possibility to pose the limbs of the phantom. The drawback of homogenous phantoms is that they do not model the layered structure of the human body, such as the skin-fat-muscle interface; and that the physical versions of these phantoms, by necessity, include a surrounding plastic shell. See Figure 9.2.

The simplest phantoms are spheres and cylinders, which are cheap to build and easy to reproduce. The simple geometrical forms are also useful for verification purposes, since Maxwell's equations can often be solved analytically for these simple cases [20, 22].

The MICS standard defines a physical phantom to be used. This phantom is an acrylic plastic cylinder with a diameter of 30 cm. The wall thickness should be

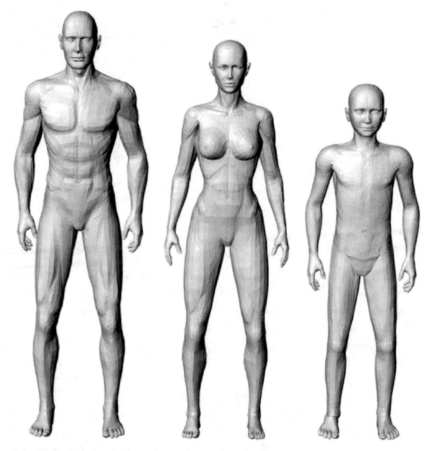

Figure 9.2   Male, female, and boy phantoms used in the simulations.

0.635 cm. It is to be filled with tissue-simulating liquid to a height of 76 cm. The medical implant should be placed on a plastic grating at a height of 38 cm inside the cylinder, at a distance of 6 cm from the sidewall. Any flexible antenna from the implant should be placed along the wall at the same height and distance. Other wires should be coiled and placed adjacent to the implant. One interpretation of the standard is that the implant is placed on the grid in the same orientation as it would be in a human torso (e.g., the pacemaker model is placed standing on its edge).

The advantage of using the MICS phantom is that it is easy to build, manage, and use. The drawback is that it is not very anthropomorphic. It resembles the chest of a human, but it has a constant curvature, in contrast to the human body, which is mostly flat on the front and back sides. One consequence of drawback is that a flat implant will be closer to the surface of the phantom at the edges of the implant, whereas the same implant in a human body would have the same distance to the skin along the whole side that is closest to the skin. The difference is illustrated in Figure 9.3.

The specification that the implant should be placed 6 cm from the sidewall of the phantom reduces this problem, but introduces a discrepancy between the placement in the phantom and the placement in an actual patient. In the case of pacemakers, the implant is most often placed subcutaneously between the fat and the pectoral

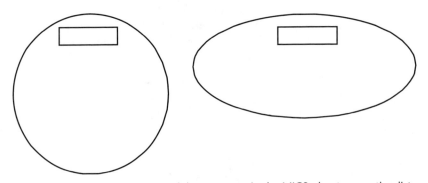

Figure 9.3 Illustration of the influence of the curvature in the MICS phantom on the distance to the edge of the phantom.

muscle beneath the collarbone. This gives an implantation depth between 0.5 and 8 cm, depending on the patient [personal communication with Dr. T. D. Fähreus, Lund University Hospital, 2002]. In the phantom, the implant is placed deeper, which introduces a larger loss to the signal due to the lossy nature of the tissue-simulating liquid. It might be that all the actual implanted cases would have a higher EIRP than is measured in the type approval procedure. Another drawback with the specified MICS phantom is that it only roughly models the chest of an adult male human, and does not represent the female and child anatomy.

There are medical implants placed at other positions in the body that also can benefit from an RF communication link. Examples are cochlea implants, which are typically mounted on the skull subcutaneously above the ear, and myoelectric sensors for control of prostheses, which are placed inside the residual muscles controlling the missing limb [6]. The existing MICS phantom models these other implantation sites very poorly, and will give erroneous results for the EIRP.

It is important to remember that the MICS phantom does not incorporate any fat or skin layers. Since the electromagnetic properties of fat are very different from those of muscle and skin, the thickness of the fat layer in a human body will influence the properties of a subcutaneously placed antenna.

9.2.4 Skin Depth

At low carrier frequencies, the magnetic field is more or less unaffected by the case of the implant and by the body. The field can thus pass through the case of a pacemaker, so that the coil of the pacemaker can be mounted inside the case. The attenuation in the case is related to the skin depth in the material. The skin depth is the depth at which the electric field has been attenuated by a factor of e^{-1} or 0.368. The explicit expression for the skin depth in a homogenous half-space is

$$\delta = \sqrt{\frac{2}{\omega\mu\sigma_e}} \tag{9.8}$$

where σ_e is the conductivity of the material, and μ is the permeability. Equation (9.8) is only valid for good conductors, where $\sigma/\omega\varepsilon \gg 1$. This is not true for all materials in the human body (e.g., fat), and in that case the skin depth is obtained from the attenuation $e^{-\alpha z}$, where α is the attenuation constant given as the real part of the propagation constant γ. The general form of the propagation constant γ is

$$\gamma = \alpha + j\beta = j\omega\sqrt{\mu\varepsilon_e}\left(1 + \frac{\sigma_e}{j\omega\varepsilon_e}\right) \tag{9.9}$$

Since α is the real part of γ, the skin depth δ now becomes

$$\delta = \frac{1}{\alpha} = \frac{1}{\text{Re}[\gamma]} \tag{9.10}$$

Equation (9.10) is easily solved numerically, and the results for some materials are given in Table 9.4. The permeability of vacuum $\mu_0 = 4\pi \cdot 10^{-7}$ Vs/Am is valid for all of the materials presented here.

9.2.5 Wave Propagation: One-Dimensional FDTD Simulations

The simplest model of the human body is a block of muscle tissue with a certain thickness, and extending to infinity in the other two dimensions. By this simplification, we are able to simulate the influence of tissues such as skin, fat, and muscle, using an efficient one-dimensional FDTD analysis [24], which illustrates the basic behavior of reflection and transmission of an electromagnetic wave in a body. The results in this section are for the MICS midband frequency of 403.5 MHz, and the corresponding tissue parameters used. For the one-dimensional case, one can also make analytical investigations, as can be found in [25]. These give the same results as the FDTD simulations.

The interesting phenomena to investigate are the behavior of the electric and the magnetic components of the electromagnetic field when a plane wave meets the body. Figure 9.4 shows the magnitude of the electric and the magnetic fields, normalized with the incoming plane wave amplitude. The surface of the body slab was placed at 1.000m, and the thickness of the slab was 144 mm, which is a typical thickness of a human body at the level of the fourth vertebrae, taken from [26]. Clearly visible are the well-known node of the electric field and the antinode of the magnetic field at the surface of the body [27, 28]. Due to this effect, pagers often use magnetic loop antennas oriented perpendicular to the body [28]. Inside the body block, we have a dominating propagating wave, which is attenuated due to the con-

Table 9.4 Calculated Skin Depths

Material	σ_e (S/m)	Skin Depth δ 170 kHz	403.5 MHz	2.45 GHz
Copper	5.8×10	$280\,\mu m$	$5.8\,\mu m$	$2.4\,\mu m$
Titanium	2.3×10	$800\,\mu m$	$16\,\mu m$	$6.7\,\mu m$
Water	[23]	13m	0.87m	0.024m
Seawater	$\sigma_{\{DC\}} = 5$ [23]	0.6m	0.013m	0.007m
Muscle Tissue	0.37/0.79/1.74 [16]	2.2m	0.052m	0.022m

The values for distilled water, seawater, and muscle tissue are found in the references given in the table.

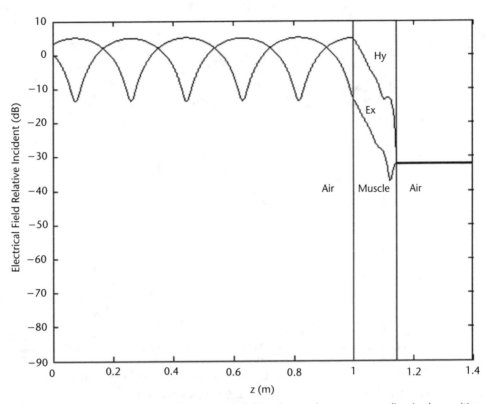

Figure 9.4   The RMS electric and the magnetic fields when a plane wave traveling in the positive z-direction hits upon a simple one-dimensional phantom.

ductivity of the muscle tissue. The magnetic field is strengthened at the surface between the body block and the air, which implies that a magnetic antenna would be beneficial also for subcutaneous implants.

If we include the skin, and the fat layer between the skin and the muscle layer, a more complex effect will be seen. Simulations done with the same body block as in Figure 9.4, but with a fat layer and a cover of 3-mm skin on each side, are shown in Figures 9.5 and 9.6. The simulations were done with fat layers of thicknesses 0, 5, 10, 25, and 50 mm. There is a variation with the thickness of the fat layer, but in these simulations, the variation is less than 2 dB at the interface between the fat layer and the muscle tissue, which is the usual placement of the medical implant.

The apparent discontinuity of the magnetic field is due to the current density in the skin

$$\hat{n} \times \left(\vec{H}_{air} - \vec{H}_{fat} \right) = \vec{J}_{skin} \cdot d_{skin} \tag{9.11}$$

where

$$\vec{J}_{skin} = \sigma_{skin} \cdot \vec{E}_{skin} \tag{9.12}$$

is the current density in the skin, and d_{skin} is the thickness of the skin. This is a fairly good approximation, since $d_{skin} << \lambda_{skin}$.

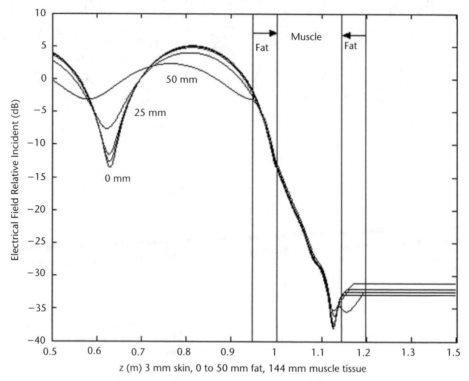

Figure 9.5 Electrical field strength dependence on fat-layer thickness.

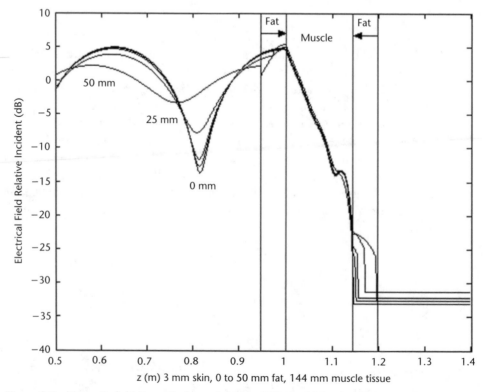

Figure 9.6 Magnetic field strength dependence on fat-layer thickness.

Simulations for the frequency of 2.45 GHz, with a homogenous muscle layer, a 3-mm outer skin layer, and a 5-mm fat layer, are shown in Figures 9.7 and 9.8, together with the corresponding results for simulations at 403.5 MHz. The amplitudes of the electric fields in Figure 9.8 are comparable for the two frequencies at the point of a subcutaneous implant (i.e., at the 1.0-m mark). This indicates that the higher attenuation of the higher frequency is not critical for this kind of implant. Since the higher frequency has a higher attenuation when penetrating the body tissue, the lower frequency is better for implants placed deeper inside the body, if we assume the same antenna performance for the implanted antenna.

9.2.6 Influence of Patient

When we place a medical implant with an antenna inside a patient, the antenna will behave differently if placed in an arm, deep in the abdomen, or just beneath the skin in the chest. There will also be a dependency on the surrounding tissue type (e.g., variations in the subcutaneous fat layer). This layer varies in thickness between patients, and varies over time, when a patient gains or loses weight. Movements of the patient change the immediate surrounding of the implanted antenna. The MICS frequency band corresponds to a wavelength of approximately 74 cm in air and approximately 9 cm inside the body. The body surface is in the near field of the implanted antenna. Thus, a change in posture changes the far-field pattern, and affects the radio channel between the medical implant and the external base station. This variation of the radio channel corresponds to a kind of slow fading. This effect can be demonstrated both by simulations [22] and measurements [19].

The gain, directivity, and efficiency of an antenna have standard definitions [29]. The antenna has an efficiency factor η, $0 \le \eta \le 1$, which is a measure of the power lost in the antenna and in the body. The ideal lossless antenna, with $\eta = 1$, is an impossibility for an implant, due to losses in the tissue. The gain definition assumes a perfectly matched antenna. Thus, reflection losses due to mismatch are not included in the gain. Gain is determined in free space in the far zone of the antenna, which in the implant case corresponds to the far zone of the body with the implant.

9.2.7 Phantom Influence on Antenna

The circumference antenna described in this chapter was inserted into the male, woman, boy, and baby numerical phantoms, as shown in Figure 9.9. The phantoms were homogenous, anatomically-based bodies. They were simulated as filled with a substance with the same characteristics as the muscle-simulating liquid. The legs and arms beneath the elbow of the phantoms were not included, in order to reduce the simulation time. The size of the baby phantom was small enough to enable the phantom to be complete. The pacemaker was implanted at approximately the same place in the male, female, and boy phantoms—subcutaneously below the left clavicle. In the baby phantom, the implant was placed in the abdomen. The size of the implant made it impossible to be placed below the left clavicle of the baby phantom. The lower abdomen is used as an implantation spot for pacemakers in small children [3]. The implants were placed at least 10 mm from the surface of the phan-

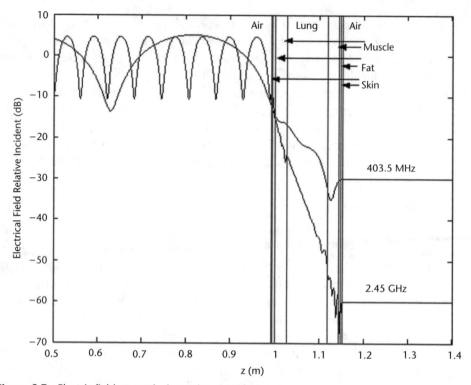

Figure 9.7   Electric field strength dependence on frequency.

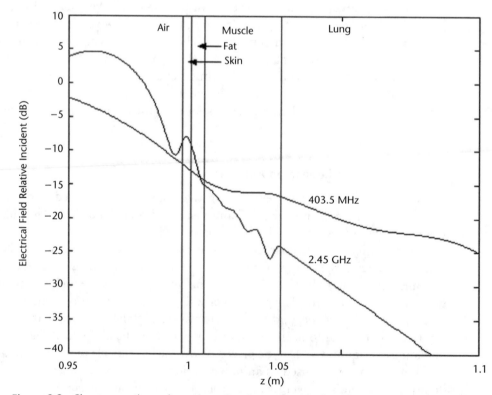

Figure 9.8   Close-up on the surface where the plane wave is reflected.

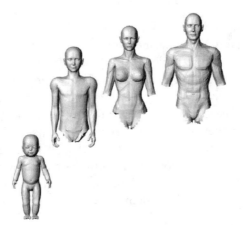

Figure 9.9   The reduced phantoms used in the simulations with the circumference antenna.

toms. The scaled heights of the phantoms are given in Table 9.5. From the simulation results in Table 9.6, we see that the efficiency is dependent on the shape of the phantom. The far-field radiation pattern is also dependent on the size and shape of the phantom.

9.3   Low-Profile Antennas for Implantable Medical Devices

Antenna design is a mature science, and an engineering discipline with a large number of design manuals available [29–31]. These books have one thing in common: they mainly describe antennas placed in a nonconducting surrounding with a relative permittivity of 1, or close to 1. In other words, they describe antennas placed in vacuum or air. The only structure that is typically found close to the antenna is a radome, which is made of low-loss materials with low permittivity. When the antenna is placed inside a human body, we have a completely different situation. The antenna is surrounded by a lossy material with high permittivity. There

Table 9.5   Height of the Phantoms, Including Legs

Phantom	Full Height	Phantom Reductions
Baby	75 cm	

Table 9.6   Maximum Gain and Radiation Efficiency from the Circumference Antenna Implanted in the Phantoms, Where the Gain Difference Is Relative the Straight Wire Antenna

Phantom	Maximum Gain	Radiation Efficiency
Baby	−25.5 dBi	$1.31 \cdot 10^{-3}$
Boy	−23.0 dBi	$1.83 \cdot 10^{-3}$
Woman	−23.6 dBi	$1.36 \cdot 10^{-3}$
Man	−26.3 dBi	$0.78 \cdot 10^{-3}$

are two instances in classical antenna applications where similar conditions occur: buried antennas and submarine antennas. King and Smith sum up this field in [25].

There are some examples of the design of high-frequency implanted antennas in the literature [20, 22, 32], and patents have been granted [33–35]. If we expand the field to "biomedical telemetry," there is much more published material. The systems described in classic texts, such as Mackay [36] and Caceres [37], use mainly coil antennas, since they use low frequencies for transmission. Most of the commercially available implantable systems today, such as Advanced Telemetry Systems [38], use coil antennas, although some transmitters use wire antennas, similar to the trailing wire antennas for submarines. The wire antennas are often used for aqueous animals, but subcutaneously implanted wire antennas are also used for birds.

9.3.1 What Is the Antenna?

When we look at the antenna implanted in a lossy and finite body, the definition of the extent of the antenna needs to be discussed. The basic definition of the antenna includes only what is attached to the implant. This disregards the influence of the implants on the antenna characteristics. Furthermore, the analysis of the radio link must include the wave propagating from the antenna through the body, into the air, and to the base station antenna. This propagation is hard to characterize, especially since it is hard to characterize the radiation pattern from the implant itself. The radiation characteristics are influenced by the tissues in the near field of the antenna, and vary between different patients.

If we now look at the entire system, we can define the implant antenna characteristics as the total of the implant antenna, the implant itself, and the body. This is the radiating structure when the implant is in place, of which we can measure the gain and the efficiency. The body shape and placement of the implant complicate the measurements. The gain, directivity, and efficiency will vary with the patient. These variations must be taken into account by adding them to the system design and to link budget calculations.

9.3.2 Antenna Efficiency Calculations in Matter

The definition of the efficiency of an antenna inside an infinite mass of lossy matter is not obvious, since the far field is attenuated to zero due to the losses. The standard definition of antenna gain is $G(\theta, \phi) = \eta D(\theta, \phi)$, where η is the efficiency factor [29]. $D(\theta, \phi)$ is the directivity of the antenna, and is defined from the normalized power pattern P_n as

$$D(\theta,\phi) = \frac{P_n(\theta,\phi)}{P_n(\theta,\phi)_{average}} = \frac{\left|\vec{F}(\theta,\phi)\right|^2}{\left|\vec{F}(\theta,\phi)\right|^2_{average}} \qquad (9.13)$$

where $\left|\vec{F}(\theta,\phi)\right|$ is the far-field amplitude.

The normalized power pattern P_n is defined from the Poynting vector, $S = \vec{S} \cdot \hat{r}$, as

$$P_n = \frac{S(\theta,\phi)}{S(\theta,\phi)_{\text{max}}} = \frac{\left|\vec{F}(\theta,\phi)\right|^2}{\left|\vec{F}(\theta,\phi)\right|_{\text{max}}^2} \tag{9.14}$$

This definition also applies to antennas inside lossy media. The normal intuitive definition of the pattern, as something measured in the (extreme) far field, does not apply. This is due to the fact that the radiating power of an antenna inside an infinite lossy medium is attenuated by the matter as it propagates outward.

The far-field amplitude is defined as

$$\vec{F}(\theta,\phi) = \lim_{|k|r \to \infty} \vec{E}(r,\theta,\phi)kre^{jkr} \tag{9.15}$$

where k is the complex wave number

$$k = \omega\sqrt{\mu\varepsilon_c} \tag{9.16}$$

The definition of the efficiency of an antenna inside a lossy medium is derived from the definition of efficiency of an antenna used in air. The usual way of defining antenna efficiency is

$$\eta_{\text{lossless}} = \frac{P_{\text{radiated}}}{P_{\text{accepted}}} \tag{9.17}$$

Here, P_{accepted} is the power that is accepted by the antenna, that is, the input power to the antenna subtracted by the reflected power from the antenna. In the case of an antenna in matter, we have to modify this definition, since the quantity P_{radiated} will vary with the radius r. The radiated power has the r-dependence

$$P(r) = P_0 e^{-2\,\text{Im}[k]r} \tag{9.18}$$

where r is the radius at which we calculate the power.

We now define the efficiency of an antenna in a lossy matter as

$$\eta_{\text{lossy}} = \frac{P_0}{P_{\text{accepted}}} \tag{9.19}$$

This definition is valid also for a lossless medium, and thus we can use the notation η for both lossy and lossless media. When the gain is given without any direction stated, the maximum gain is implied. The same applies to the directivity. The gain of antennas, in both air and in matter, is thus defined as

$$G = \eta D \tag{9.20}$$

In order to measure or calculate P_0, we use

$$P_0 = \text{Re}\int_S \vec{S}(r,\theta,\phi)e^{2\,\text{Im}[k]r} \cdot \hat{r}dS \tag{9.21}$$

The surface S is a sphere in the far zone of the antenna with the center at the origin. Notice that P_0 depends on the location of the origin when the medium is lossy. $\bar{S}(r,\theta,\phi)$ is the complex Poynting vector

$$\bar{S}(r,\theta,\phi) = \frac{1}{2}\bar{E}(r,\theta,\phi) \times \bar{H}^*(r,\theta,\phi) \qquad (9.22)$$

The problem of measuring the efficiency of the antenna in a homogenous medium is not always important for an implant application, since usually we are interested only in systems that communicate from the inside of a lossy medium to a device outside in air. Thus, the relevant measurements of the system include the lossy body, and efficiency measurements can utilize a phantom. However, in the development of different antennas, it is practical to be able to compare them by efficiency, especially since there is a large risk of the accepted power being absorbed by the surrounding lossy liquid in the near field. This makes the far-field measurements less useful. Care must still be taken with the air-skin-fat layer interface that is part of the surroundings of a subcutaneous implant, which is not part of these calculations.

9.3.3 Electric Versus Magnetic Antennas

When we place an antenna in matter, there are a number of changes in comparison with the antenna placed in air. The main difference is that the wavelength changes, which is due to the change in ε_e and σ_e, [see (9.4) and (9.5)]. The wavelength in the material is shorter, since the wave propagation speed is lowered. The reduction becomes

$$\lambda_m = \frac{\lambda_0}{\mathrm{Re}\left[\sqrt{\varepsilon_{er} - j\dfrac{\sigma_e}{\omega\varepsilon_0}}\right]} \qquad (9.23)$$

Losses in the material will also affect both the near field and the wave propagation. The electromagnetic field from a small antenna in a lossy material can be expressed in terms of the currents in the antenna. By examining Maxwell's equations, the currents can be used to discuss some fundamental aspects of small antennas in matter. We assume a sphere with a radius a and a volume V. Outside the sphere, the electric field can always be expressed as a multipole expansion. The expansion reads

$$\bar{E}(\bar{r}) = \sum_{l=1}^{\infty}\sum_{m=0}^{l}\sum_{\tau=1}^{2} a_{\tau ml(\mathrm{even})}\bar{u}_{\tau ml(\mathrm{even})}(\bar{r}) + a_{\tau ml(\mathrm{odd})}\bar{u}_{\tau ml(\mathrm{odd})}(\bar{r}) \qquad (9.24)$$

which follows from an expansion of the Green function in spherical waves [39, 40]. Every term in the sum constitutes an outwardly propagating spherical wave, called a partial wave. The explicit expressions for the partial waves $\bar{u}_{\tau ml}(\bar{r})$ are given in [22]. We can obtain any radiation pattern by a suitable set of partial wave amplitudes $a_{\tau ml}$. This can be achieved by designing the currents in the volume V that give this set. There is no limit to the size of the volume V (i.e., the antenna can be arbitrarily small). The l-value is linked to the angular variation of the field. In [39, 40], it is shown that the optimal directivity of an antenna with a maximum index l_{max} is bounded by

$$D \leq l_{max}(l_{max} + 2) \tag{9.25}$$

Equation (9.25) shows that in order to get a large directivity D, we need to have a large l. From the derivations in [22], we see that in the near-zone, $|kr| \ll 1$, and the partial wave of index l is proportional to

$$\vec{u}_{\tau m l}(\vec{r}) \sim \begin{cases} (kr)^{-l-1} ; \tau = 1 \\ (kr)^{-l-2} ; \tau = 2 \end{cases} \tag{9.26}$$

The corresponding power flow in a lossless material is proportional to $(kr)^{-2l-3}$, which shows that the near-field grows rapidly with l. This implies that the electric and magnetic energies that are stored in the near-zone will grow rapidly with l. The stored energy is linked to a reactive power flow, and does not contribute to the radiation from the antenna. To get a large directivity, we need a large l-value. This gives large reactive near-fields around the antenna, which implies large nonradiating currents in the antenna. Since the metal in the antenna has a finite conductivity, the nonradiating currents give rise to ohmic losses. This, in turn, leads to the low gain of the high-directivity small antenna.

This result still holds when we add losses to the matter into which the antenna is placed. The large near-fields will then be an even larger problem. Since the surrounding matter is lossy, the wave number k is complex, and the reactive fields are not purely reactive and will lose energy to the matter. This power loss is nonradiating, since it consists of ohmic losses in the near field of the antenna (i.e., heat). The antenna in a lossy matter thus loses power in three ways: ohmic losses in the antenna, ohmic losses from the near field in the matter, and radiated power. The radiated power will be attenuated by the lossy matter and converted to heat as it propagates. The accepted power in the calculation of the radiation efficiency then reads

$$P_{accepted} = P_{ohm} + P_{near-field} + P_0 e^{-2\,Im[ka]} \tag{9.27}$$

where a is the radius of a lossless sphere in which the antenna is confined. The radiated power loss is independent of l, while the other two losses increase with l. Thus, in the case of antennas in a large body of lossy matter, one should keep the l-value low, which gives a low-directivity antenna. Using an insulator around the antenna can reduce the power loss in the near zone.

This is illustrated by the graphs in Figure 9.10, which are calculated by numerical evaluation of the multipole expansions in muscle tissue at 400 MHz. Figure 9.10 illustrates that dipole antennas are more efficient than higher order antennas, and that magnetic antennas are more efficient than electric ones. It also illustrates that the efficiency of the antenna increases by increasing the insulation surrounding the antenna. A magnetic dipole antenna should have at least 2 mm of insulation, and an electric antenna should have from 4 to 6 mm of insulation.

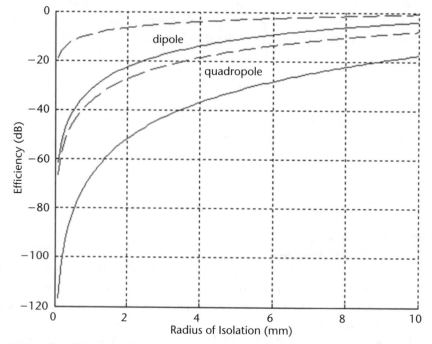

Figure 9.10 The efficiency of an antenna inside a lossless sphere of radius *a* inside muscle tissue. Solid lines are for electric antennas and dashed lines for magnetic antennas. Dipoles have *l* = 1, and quadropoles have *l* = 2. (*After:* [39].)

Dipole antennas only create partial waves with l_{max} = 1. The maximum directivity of such an antenna is 3, according to (9.25). There are three main choices of dipoles: the electric dipole, the magnetic dipole, and the combined dipole. The magnetic dipole is the most power-efficient antenna, if we do not take into account the ohmic losses in the antenna. It is typically a coil that is resonated at the correct frequency by an external capacitor to get a resistive antenna impedance. The directivity of the magnetic antenna is 1.5, or 1.8 dB [29]. The electric dipole is less power-efficient, since the electric near field is stronger. Lower ohmic losses in the antenna often compensate for this reduced power efficiency, since the current strengths are lower. The electric dipole can be made resonant by connecting an inductor in series with the antenna. The directivity of the electric dipole is the same as for the magnetic dipole (i.e., 1.5). The most common type of dipole is the combination of an electric and a magnetic dipole. This can be made resonant in itself by a proper combination of the two dipoles. The common resonant half-wave dipole is an example of this type of antenna. Placed in free space, it is efficient since both the magnetic and the electric components radiate, keeping the nonradiating currents low. A directivity of 3 can only be achieved when the electric and magnetic dipole moments are perpendicular to each other.

9.3.4 Implantable Antenna Designs

There are practically an infinite number of different designs that could be used for implantable antennas. The difference between implantable antennas and antennas in air is that only a few of the implantable antennas have been tested and documented.

Two main groups of implantable antennas are the external wire antenna, and antennas that conform to the shape of the implant. The classic dipole antenna has also been used for diverse theoretical and basic studies, but it is not well suited for implantation use, since it require a large stiff extension from the implant.

We will now discuss some examples of implant antennas. These include:

- Trailing wire antennas;
- Circumferential wire/PIFA antennas;
- Microstrip antennas;
- Patch antennas;
- Magnetic coil antennas.

9.3.4.1 Wire Antenna

One of the antennas used for implants is the wire antenna [38, 41]. The basic function is the same as the classical long-wire antenna [42], with some differences. Since the medium surrounding a wire antenna in matter is lossy, the traveling wave is attenuated as it travels along the wire. When the wave is reflected at the end of the wire antenna, it will travel back towards the feedpoint. Thus, the impedance at the feedpoint depends on the length of the antenna, and on the reflection at the furthest end. The uninsulated bare wire antenna in a medium with $\varepsilon_r \neq 1$ and $\sigma \neq 0$ is the simplest structure. The phase velocity of the electromagnetic wave in the wire is the same as the phase velocity in the medium outside the wire: $v_p = v_c$, where v_p is the phase velocity in the wire antenna, and v_c is the phase velocity in the matter surrounding the antenna. This makes the antenna a so-called slow wave structure [29].

The phase velocity in the medium is

$$v_c = \frac{c_0}{\sqrt{\varepsilon_{er}}} \qquad (9.28)$$

where ε_{er} is defined in (9.6).

Figure 9.11 shows the instantaneous magnitude of the electric field around a wire in a lossy medium with a conductivity $\sigma_e = 0.9$ S/m. The simulated antenna is 360 mm long, and is fed with a 403.5-MHz signal. The wavefronts are circular, and meet the wire at 90 degrees. The wave is reflected at the end of the wire antenna and forms a standing wave pattern. The amplitude of the field is attenuated along the wire, and the standing wave pattern in the wire is much less pronounced than would be the case in a lossless medium. The impedance of a long wire antenna is independent of the exact length of the wire.

If we surround the wire antenna with an insulation that has a much lower permittivity than the surrounding matter, then we alter the phase velocity in the wire. The loss per unit length is lower, since the lossy matter is now removed from the region of the strongest near-field. Thus, the reflection has a larger impact on the

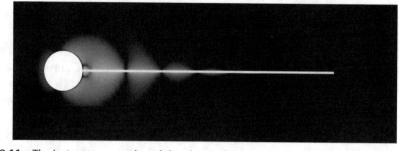

Figure 9.11 The instantaneous value of the electric field around a wire antenna in muscle tissue, with ε_{er} = 62.5, and σ_e = 0.9. The simulated antenna is 360 mm long, and is fed with a 403.5-MHz signal.

impedance. This antenna compromises a fast-wave structure, with $v_p > v_c$. This is clearly seen in Figure 9.12, which illustrates the fields around an isolated wire in a lossy matter. Here, the wavefronts are at an angle of less than 90 degrees from the wire, since the phase front moves faster in the wire than in the lossy matter. The fast-wave structure is common in leaky wave designs of antennas [29]. Due to the reduced loss, the reflected wave has a nonnegligible amplitude when it reaches the feedpoint, and we have a standing wave pattern along the antenna.

In a lossy matter, the insulated wire antenna can be treated as a coaxial waveguide [25]. The lossy matter acts as the outer conductor. This waveguide has a propagation constant γ, which is influenced both by the dielectric properties of the insulation and by the properties of the surrounding matter. In [43], an approximate solution to the input impedance of the insulated wire antenna is presented. The impedance of the wire antenna is

$$Z_{\text{wire}} = Z_0' \coth(\gamma l) \tag{9.29}$$

where l is the length of the wire, and γ is the complex propagation constant approximated by

$$\gamma \approx \sqrt{-\omega^2 \mu_2 \varepsilon_2} \left(1 - \frac{(j\pi/4) + \ln\left(0.89\sqrt{2}\,\dfrac{a_{\text{out}}}{\delta}\right)}{\ln\dfrac{a_{\text{out}}}{a_{\text{in}}}} \right)^{1/2} \tag{9.30}$$

Figure 9.12 The instantaneous electric field around an isolated wire antenna in muscle tissue, with ε_{er} = 62.5, and σ_e = 0.9. The simulated antenna is 360 mm long, and is fed with a 403.5-MHz signal.

Here, μ_2 and ε_2 are the electromagnetic properties of the insulation, a_{out} and a_{in} are the outer and inner radii of the insulator, respectively, and δ is the skin depth in the surrounding material. The characteristic impedance Z'_0 is approximated in [43] by

$$Z'_0 = \left(\frac{1}{2\pi} \ln \frac{a_{out}}{a_{in}} \right) \frac{\gamma}{j\omega\varepsilon_2} \tag{9.31}$$

and the skin depth is given in (9.10). The approximation in (9.30) is valid if the propagation constant for the lossy outer medium is much greater than the line propagation constant γ. The approximation is valid if $a_{out}/\delta < 0.1$.

For a wire antenna inserted into the muscle-simulating liquid, with the wire dimensions of a_{out} = 1.5 mm, a_{in} = 1 mm, and l = 0.36m, and with ε_2 = 2 at f = 403.5 MHz, (9.29) gives an antenna impedance of

$$Z_{wire} = 54 - j12\Omega \tag{9.32}$$

In practice, these calculations show good agreement with simulations and measurements, as illustrated in Table 9.7.

9.3.4.2 Circumference Antenna

The circumference antenna is a compact antenna that is mounted around the edge of the pacemaker case. The circumference antenna is an interesting antenna, since it conforms to the shape of the medical implant onto which it is placed. Thus, it does not influence the mechanical properties of the implant to a large degree. The analysis of the circumference antenna follows from the monopole quarter-wave antenna in air. From [44], we get the relation between an antenna in air and the same antenna in a material with complex μ_c and ε_c as

$$\frac{1}{Z_{in}} Z(\omega, \varepsilon_c, \mu_c) = \frac{1}{Z_0} Z\left(\sqrt{\frac{\mu_c \varepsilon_c}{\mu_0 \varepsilon_0}} \omega, \varepsilon_0, \mu_0 \right) \tag{9.33}$$

In the lossy case, the resonance angular frequency becomes complex, corresponding to a damped resonance. The drawback of the circumference antenna is that it has a very low impedance, in the range from 5Ω to 10Ω. This is not necessarily bad for the transmitting case, but complicates the construction and testing, since the universal standard measurement impedance is 50Ω. It is possible to increase the bandwidth of this antenna by replacing the wire with a thin plate. See Figure 9.13.

Table 9.7 Measured Impedances of the Wire Antennas

	Re[Z]	Im[Z]
Uninsulated wire antenna	30	3
Insulated wire antenna	50	−8

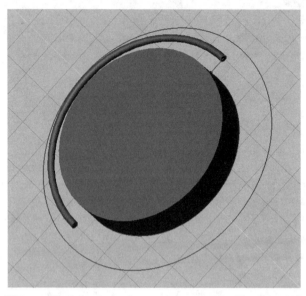

Figure 9.13 The CAD model of the circumference wire antenna. The thin circle is the radial extent of the isolation.

9.3.4.3 Circumference PIFA

The circumference planar inverted-F antenna (PIFA) is an adaptation to the circumference geometry of the inverted-F antenna (IFA) [28]. We implemented it by modifying the circumference plate antenna described above. The result is depicted in Figure 9.14. The added part of the antenna will behave as if an inductance to ground was added at the feedpoint. This antenna has the benefit of having an impedance approximately equal to 50Ω, and is thus easy to match to the classical RF impedance

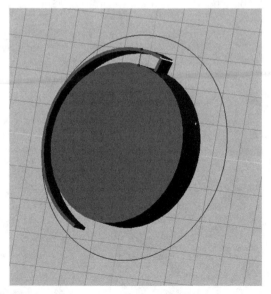

Figure 9.14 The CAD drawing of the circumference PIFA antenna. The thin circle is the radial extent of the isolation.

standard of $Z = 50\Omega$. The bandwidth is large compared with the circumference antenna. This antenna is closely related to the microstrip PIFA antenna described next.

9.3.4.4 Microstrip Antenna and PIFA

An antenna can be placed on the large flat surface of a typical implant. This calls for a flat antenna, typically of the microstrip type. These antennas typically consist of a flat slab of dielectric substrate, with one side covered with metal, which makes up the ground plane, and the other side built up as a flat metal structure. A typical example can be seen in Figure 9.15 [20]. These are well-suited for implant use, since they are flat and conform to the case of an implant. The drawback is that the conductor on the surface will be in direct contact with the lossy medium, giving rise to large losses. The solution is to cover the microstrip side with another layer of dielectric material, which will increase the volume of the antenna. One important aspect is that, in classical antenna analysis, the use of a dielectric will produce a physically smaller antenna, since the permittivity of the dielectric is higher than air. Inside the body, a typical high permittivity material, with $\varepsilon_r = 10$, has a lower permittivity than muscle or skin tissue, with $\varepsilon_r = 45$ to 60, and is approximately the same as the permittivity of fat, with $\varepsilon_{r(403\ \text{MHz})} = 6$. Due to this, the use of a dielectric isolation on the microstrip antenna will give a larger antenna, compared to an antenna placed directly in the tissue itself.

Simulations in [20] show that the PIFA antenna has better efficiency than the microstrip antenna. This is explained by the fact that the PIFA antenna utilizes both electric fields and high currents in the loop created by the grounding pin. One-half of the antenna will thus work as a magnetic antenna. This magnetic monopole has lower losses than the electric monopole, as shown above, which explains the higher efficiency.

9.3.4.5 Patch Antenna

The patch antenna seems to be the perfect match for the medical implant. It is compact, small compared with the wavelength, and can be placed on the flat side of the implant. Its usefulness is debated, as it is difficult to make it to work properly inside

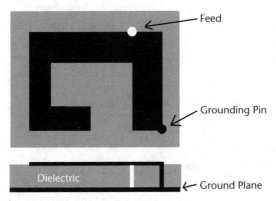

Figure 9.15 The microstrip PIFA antenna.

a lossy material. The patch antenna is a resonant structure, with a plate placed over a ground plane. Because of the plate's dimensions, there will be a standing wave across the patch. It is fed at a position that excites the resonating modes, and has a matching impedance to the antenna feed. The space between the patch and the ground plane is often loaded with a dielectric substrate in order to reduce the wavelength and the size of the patch.

If we place a patch antenna in a lossy matter, the reflections at the ends of the patch will be reduced, since the patch surface is essentially extended infinitely by the lossy matter. The wave propagating between the patch and the ground plane is a surface wave. It is not reflected at the edges of the patch, but continues to propagate as an attenuated surface wave. The radiation from this is very small. We thus lose the resonant structure of the patch.

If we cover the antenna with a thick insulation, we get a useful antenna, but this covering must be of a very high epsilon material in order to keep the patch sufficiently small to be placed on an implant.

9.3.4.6 Coil Antenna

As shown earlier, the magnetic field is increased at the interface between the air and the body, and an antenna operating as a magnetic antenna should then be advantageous. They are also a good choice, based on the theoretical efficiency calculations of antennas in lossy matter. Ferrites are often used to improve the performance of magnetic antennas. The ferrite material has an upper frequency limit, over which the permeability of the material falls. Typical ferrite materials do not work at frequencies as high as 400 MHz.

A magnetic coil antenna is inductive at low frequencies. There is a capacitive coupling between adjacent turns of the coil. Thus, each turn constitutes a resonant circuit. The proportionality of the self-resonance frequency of each turn of the coil is defined as

$$f_r \propto \frac{1}{2\pi \sqrt{L_e C_t}} \tag{9.34}$$

where C_t and L_e are the capacitance and the inductance in a loop, respectively [45]. At frequencies lower than f_r, the coil is inductive, and at frequencies higher than f_r, the coil is capacitive. In order to have an effective magnetic antenna, it should be used at a frequency below the self-resonance. In order to get an efficient antenna, the radiation resistance R_{rad} should be as high as possible. It is possible to show that the radius should also be made as large as possible.

9.3.5 Dependence on Insulation Thickness

The thickness of a nonconducting insulation has a major impact on the antenna performance of all antenna types. To illustrate this, the circumference PIFA antenna was simulated with different thicknesses of the radial insulation. The result is shown in Figure 9.16. As can be expected, the resonant frequency of the antennas was lowered, due to an increased effective permittivity in the medium surrounding the

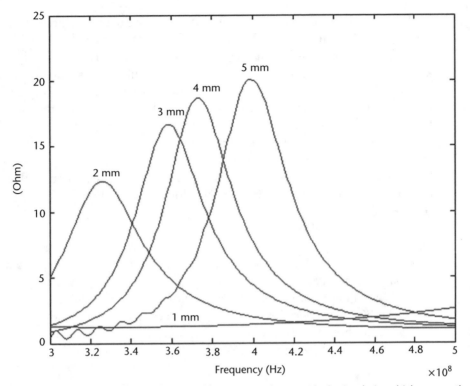

Figure 9.16   Variation of the real part of the PIFA impedance with the insulating thickness on the outside of the antenna.

antenna. The radiation efficiency changed for the different antennas, as can be seen in Table 9.8. This shows that any amount of isolation put on the PIFA antenna is beneficial. If the isolation is reduced below 3 mm, the efficiency falls drastically.

9.3.6   SAR

The amount of heat that is generated in the tissue surrounding the implant antenna is of vital importance for the carrier of the implant. This is measured as SAR, in watts per kilogram. It is typically taken as an average value over a certain cubic volume of mass. Various countries have different regulations for the calculation of the

Table 9.8   Center Frequency and Efficiency Variation with Covering Isolation Thickness in the PIFA Antenna

Thickness	Frequency	Efficiency
5 mm	403.5 MHz	100%
4 mm	373 MHz	92%
3 mm	358 MHz	83%
2 mm	321 MHz	62%
1 mm	233 MHz	12%

The efficiencies are given relative to the case with 5 mm of insulation thickness.

SAR and its limits. Two typical limits are: 1.6 W/kg over a volume of 1g, as established by the FCC for the United States; and 2 W/kg over a volume of 10g, as established by the European Union [46]. If we assume the density of the tissues to be 1 kg/dm³, 10g corresponds to an integration over a cube with a side of 2.15 cm and a volume of 10 cm³. The size of this volume is comparable to the volume of a typical implant. The maximum SAR values are directly dependent on the electrical field from the antenna, and thus on the size of the near field from the antenna. In order to exemplify the difference between the different averaging volumes, the SAR for an implanted antenna was calculated over three different averaging volumes in Table 9.9. The table is normalized to 1 mW input power into the antenna. The values in the table were calculated according to the draft specifications SCC-34, SC-2, WG-2, IEEE-P1529/D0.0 [47].

The SAR limit of the FCC cannot be exceeded by a transmitter with an output power of less than 1.6 mW, since even if all the power was absorbed in the same 1g cube, the limit of 1.6 W/kg would not be reached. In the same way, a transmitter of less than 20 mW cannot exceed the Swedish limit of 2 W/kg. Correspondingly, devices with a mean output power below 0.02W do not need to be tested in Sweden [46]. The FCC has a proposal [48] that devices worn at the body with an output below 2 mW should be exempt from licensing. There seem to be no existing specific regulations on SAR for implants.

9.4   Conclusion

Medical implant communication systems are a growing application field for body-centric communications. The most significant complication with these systems is that the antennas are placed inside the human body, which is a lossy object. This places special demands on the antenna design, in addition to the obvious requirements that the implant and the antenna should be made of biocompatible materials and should be nonhazardous for the patient. Theoretical investigations on implanted antennas in large objects suggest that the most efficient antenna is a pure magnetic antenna. Practical design experiments with antennas incorporating a magnetic component indicate that these antennas perform better than pure electric ones. The amount of isolation around the antenna is important for its efficiency. The complicated surroundings of a subcutaneous implant, with the air-skin-fat-muscle interface, make the interpretation of the practical implications for these theoretical results more difficult. The development and testing of implanted antennas are com-

Table 9.9   SAR Levels of the Implanted Antennas

Antenna	SAR (mW/kg)		
	10g	1g	1 mg
Uninsulated wire	52	300	7,900
Isolated wire	4.3	12	47
Circumference wire	8.1	14	150
Circumference plate	8.3	22	260
Circumference PIFA	8.0	21	240

plicated, since anthropomorphic phantoms are complex and nonstandardized. The simple phantoms used today are too simplistic to give realistic evaluations of the performance of the implant communication link when placed in an actual patient. FDTD simulations are therefore a very useful tool in implant antenna design, since the results from such simulations agree well with actual measurements.

References

[1]   Brunckhorst, C., R. Candinas, and S. Furman, "Obituary: Ake Senning, 1915–2000," *J. of Pacing and Clinical Electrophysiology*, Vol. 23, No. 11, November 2000.

[2]   Greatbatch, W., *The Making of the Pacemaker*, Amherst, NY: Prometheus Books, 2000.

[3]   Webster, J. G., (ed.), *Design of Cardiac Pacemakers*, New York: IEEE Press, 1995.

[4]   Cavuoto, J., "Neural Engineering's Image Problem," *IEEE Spectrum*, April 2004.

[5]   Medtronic, Inc., http://www.medtronic.com.

[6]   Weir, R. F., et al., "Implantable Myoelectric Sensors (IMES) for Upper-Extremity Prosthesis Control," *Engineering in Medicine and Biology Society, 2003, Proc. 25th Annual Int. Conference of the IEEE*, September 2003, pp. 1562–1565.

[7]   Woermannkehre 1, D-12359 Berlin, Germany, BIOTRONIK GmbH and Co., http://www.biotronik.com.

[8]   Scanlon, W. G., "Analysis of Tissue-Coupled Antennas for UHF Intra-Body Communications," *IEE ICAP*, January 2005.

[9]   Finkenzeller, K., *RFID Handbook*, 1st ed., New York: John Wiley & Sons, 1999.

[10] De Balthasar, C., et al., "Design of Antennas to Power Injectable Micro-Stimulators: A Systematic Approach," *9th Annual Conference of the Int. FES Society*, August 2004.

[11] European Telecommunication Standards Institute, http://www.etsi.org.

[12] ETSI EN 301 839-1 Electromagnetic Compatibility and Radio Spectrum Matters (ERM); Radio Equipment in the Frequency Range 402 MHz to 405 MHz for Ultra Low Power Active Medical Implants and Accessories; Part 1: Technical Characteristics, Including Electromagnetic Compatibility Requirements, and Test Methods, European Telecommunications Standards Institute, 2002.

[13] 47 CFR 95.601-95.673 Subpart E, Federal Communications Commission, 1999.

[14] Recommendation ITU-R SA.1346, International Telecommunication Union, 1998.

[15] "Planning for Medical Implant Communications Systems (MICS) and Related Devices," Australian Communications Authority, Proposals Paper SPP 6/03, October 2003.

[16] Gabriel, C., *Compilation of the Dielectric Properties of Body Tissues at RF and Microwave Frequencies*, Brooks Air Force, Technical Report AL/OETR-1996-0037, 1996.

[17] Hartsgrove, G., A. Kraszewski, and A. Surowiec, "Simulated Biological Materials for Electromagnetic Radiation Absorption Studies," *Bioelectromagnetics*, No. 8, 1987, pp. 29–365.

[18] "The Visible Human Project," United States National Library of Medicine, 2004, http://www.wma.net.

[19] Scanlon, W. G., J. B. Burns, and N. E. Evans, "Radiowave Propagation from a Tissue-Implanted Source at 418 MHz and 916.5 MHz," *IEEE Trans. on Biomedical Engineering*, April 2000, pp. 527–534.

[20] Kim, J., and Y. Rhamat-Samii, "Implanted Antennas Inside a Human Body: Simulations, Designs and Characterizations," *IEEE Trans. on Microwave Theory and Techniques*, Vol. 52, No. 8, August 2004, pp. 1934–1943.

[21] Curios Labs, http://www.e-frontier.com/go/products/poser.

[22] Johansson, A. J., *Wireless Communications with Medical Implants: Antennas and Propagation*, 1st ed., Lund, Sweden: Lund University, 2004.

[23] Buchner, R., J. Barthel, and J. Stauber, "The Dielectric Relaxation of Water Between 0C and 35C," *Chem. Phys. Letters*, Vol. 306, No. 1–2, June 1999.

[24] Taflove, A., *Computational Electrodynamics*, Norwood, MA: Artech House, 1995.

[25] King, R. W. P., and S. G. S., *Antennas in Matter*, Cambridge, MA: MIT Press, 1981.

[26] Köpf-Maier, P., *Atlas of Human Anatomy*, 5th ed., Berlin, Germany: Karger, 2000.

[27] Durney, A. H., and D. A. Christen, *Basic Introduction to Bioelectromagnetics*, Boca Raton, FL: CRC Press, 1999.

[28] Fujimoto, K., and J. R. James, (eds.), *Mobile Antenna Systems Handbook*, 2nd ed., Norwood, MA: Artech House, 2001.

[29] Balanis, C. A., *Antenna Theory*, 2nd ed., New York: John Wiley & Sons, 1982.

[30] Kraus, J. D., *Antennas*, New York: McGraw-Hill, 1988.

[31] Kildal, P.-S., *Foundations of Antennas*, Lund, Sweden: Studentlitteratur, 2000.

[32] Furse, C., "Design of an Antenna for Pacemaker Communication," *Microwaves and RF*, 2000, pp. 73–76.

[33] Sun, W., G. Haubrich, and G. Dublin, U.S. Patent No. 5,861,091, "Implantable Medical Device Microstrip Telemetry Antenna," USPTO, 1999.

[34] Amundson, M. D., et al., U.S. Patent No. 6,456,256, "Circumferential Antenna for an Implantable Medical Device," USPTO, 2002.

[35] Von Arx, J. A., et al., U.S. Patent No. 6,708,065, "Antenna for an Implantable Pacemaker," USPTO, 2004.

[36] Mackay, R. S., *Biomedical Telemetry*, 1st ed., New York: John Wiley & Sons, 1968.

[37] Caceres, C. A., (ed.), *Biomedical Telemetry*, New York: Academic Press, 1965.

[38] Advanced Telemetry Systems, Inc., 2004, http://www.atstrack.com.

[39] Karlsson, A., "Physical Limitations of Antennas in a Lossy Medium," *IEEE Trans. on Antennas and Propagation*, Vol. 52, No. 8, August 2004, pp. 2027–2033.

[40] Harrington, R. F., *Time Harmonic Electromagnetic Fields*, New York: McGraw-Hill, 1961.

[41] Mackay, R. S., *Biomedical Telemetry*, 2nd ed., New York: IEEE Press, 1993.

[42] Kraus, J. D., *Electromagnetics*, 4th ed., New York: McGraw-Hill, 1992.

[43] Fenwick, R. C., and W. L. Weeks, "Submerged Antenna Characteristics," *IEEE Trans. on Antennas and Propagation*, Vol. 11, No. 3, May 1963.

[44] Deschamps, G. A., "Impedance of an Antenna in a Conducting Medium," *IEEE Trans. on Antennas and Propagation*, Vol. 10, No. 5, September 1962.

[45] Stutzman, W., and G. Thiele, *Antenna Theory and Design*, New York: John Wiley & Sons, 1998.

[46] *Mobiltelefoner och Strålning*, Sweden's Radiation Protection Authority, Tech. Rep., 2003.

[47] "SEMCAD: Addendum to the Reference Manual," Shmid and Partner Engineering AG, Manual, 2004, http://www.semcad.com.

[48] *Proposed Changes in the Commission's Rules Regarding Human Exposure to Radiofrequency Electromagnetic Fields*, Federal Communications Commission, Tech. Rep., 2003.

CHAPTER 10

Summary and Conclusions

Yang Hao and Peter S. Hall

10.1 Overview and Conclusions

The main objective of research and development in body-centric wireless communications is to provide users with a wide range of personal electronic support systems, anywhere and at any time. To ensure the efficient performance of such wireless systems in a very variable environment, optimum antenna design and careful characterization of the radiowave propagation channel must be performed. Antenna design, channel characteristics, and radio system performance research presented in this book will assist in developing both power-efficient and spectrum-efficient body-centric wireless systems.

The book has given an overview of fundamental electromagnetic properties of the human body and associated numerical modeling techniques for body-centric antennas and radio propagation. Detailed measurement procedures have been described for all frequencies of interest, covering UHF/VHF, RF, UWB, and so forth. Table 10.1 presents a few technologies that have been covered in this book that are suitable for body-centric wireless communications. Some of them, such as BodyLAN [1] and especially Bluetooth [2], are already in wide use, while others will become available in the near future.

In this chapter, we will try to summarize the book and highlight future challenges, such as the need for continued antenna and propagation research, and the important trend toward the coexistence and/or convergence of various networks (e.g., WBANs, WSNs [3], WPANs [4], and WLANs [5]) and services, which may require the establishment of new standards and business models.

10.1.1 Overview of Narrowband Systems

The definition of narrowband or wideband systems for body-centric wireless communications is not straightforward. Here, we differentiate between the two only by their operational bandwidths. Narrowband systems have been dealt with in most of the chapters (Chapters 3–5 and Chapters 7–9) in this book, and represent current market trends in body-centric wireless communications. Such systems use a wide selection of frequency bands.

271

Table 10.1 Summary of Available Technologies in Body-Centric Wireless Communications

Standard	Frequency (MHz)	Data Rate	Maximum Power	Range (m)	Chapter Number in the Book
UHF/VHF	~10	Very low	Very low	<=0.5	4
Medical implant (licensed)	402–405	Low	Low	<=2	8, 9
BodyLAN	900	32 Kbps	0 dBm	2–10	8
Bluetooth	2,400–2,480	1 Mbps	0 dBm	0.1–10	3, 7, 8
ZigBee	2,400 915 868	250 Kbps 40 Kbps 20 Kbps	Low	1–100	3
WLAN	2,400 5,200	10–50 Mbps	0 dBm	30–50	8
UWB	3,100–10,600	1 Gbps	−41 dBm/MHz	10	5

1. *UHF/VHF Bands:* Pacemakers and implantable RFID devices (Chapter 9) use the inductive link with a carrier frequencies ranging from 9 to 315 kHz, with a data rate of up to 512 Kbps. The range of communication is effectively constrained to touch range, and thus limits its usefulness. The problem can be slightly alleviated by using a higher frequency (~10 MHz), and thus, a further separation between the transmitter and receiver. Chapter 4 proposes a waveguide transmission channel, based on surface waves propagating along the human body. The communication system uses the near-field region of the electromagnetic wave generated by the devices, which is eventually coupled to the human body via electrodes. An average of 20-dB attenuation for each electric and magnetic field component is reported at a distance of 10 cm from the signal electrode. This value is comparable to the values (Figure 3.5) presented in Chapter 3, for the case of an on-body link when the channel is static.

2. *MICS Standard Band:* The MICS band is from 402 to 405 MHz [6]. Chapters 7 and 9 presented wireless communications for medical implants and swallowable sensors, which have distinct channel behaviors. First, electric and magnetic fields experience high attenuation inside the human body, due to the increasing conductivity of human tissue with frequencies of interest. Second, channel properties cannot be experimentally characterized easily, and can only be determined by numerical modeling. Furthermore, while most current studies examine in-to-out body channels, only in-body channel characteristics are useful when the link is established among various implants in the same patient. Figure 10.1 shows the comparison of electric fields at MICS and ISM bands (i.e., 868 MHz and 2.45 GHz), in a plane along the human stomach, clearly demonstrating the lossy behavior of human tissues at different frequency bands. Since communication links are usually needed between the implants and body-worn sensors, Chapters 7 and 9 also addressed important channel characteristics from in-body to on-body links, and from in-body to off-body links. Chapters 8 and 9 discussed various antenna types for

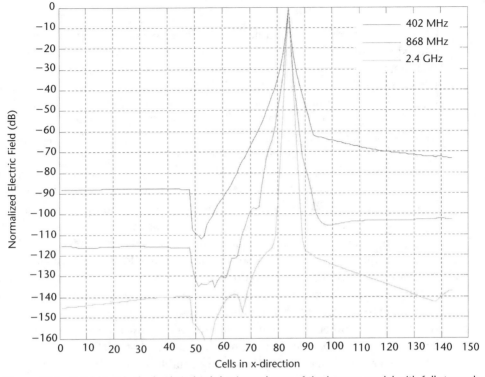

Figure 10.1   E-field inside the body in both horizontal axes of the human model with full stomach.

implantable devices, and Chapter 8 specifically demonstrated practical antenna miniaturization techniques for body sensors.

3. *ISM, Bluetooth, and WLAN:* On-body propagation measurement, models, and simulations have been presented and discussed extensively in Chapters 3, 6, and 7. It is noted that antenna properties will be modified by the human body. Communication nodes on the body are often in the near-field region. Therefore, on-body radio behavior is strongly dependent on the type antenna. Chapter 6 discussed various wearable textile antennas for space and military applications. Examples for cellular phones, GPS, and WLAN applications were presented, together with antenna performance enhancement technique using EBGs. One of the distinct features of the on-body radio channel is its variability in path loss and delay profile, due to antenna placement and posture changes. The comparison between the standing and sitting body has shown that the location of the transmitter should be chosen wisely, since its orientation can be changed by body movements. This may adversely affect the signal strength for receivers on the back. In most cases, statistical analysis of the measured on-body radio channels regarding path loss and deviation can be fitted to known distributions. The dependency of system performance on the applied data rate was demonstrated, and it was shown that acceptable BER can be achieved for power levels as low as −40 dBm for low data rates, mainly used in health-monitoring applications.

10.1.2 Overview of Wideband Systems

The wideband body-centric wireless communication system referred to in Chapter 5 is associated with UWB regulations issued by the FCC in April 2002. Under Part 15 of the rules, the FCC permits ultrawideband intentional emissions, subject to certain frequency and power limitations that will mitigate the risk of interference to those sharing the same spectrum [7]. The wideband body-centric wireless communication system can directly deploy UWB signals between 3.1 and 10.6 GHz, at power levels up to −41 dBm/MHz, with a higher degree of attenuation required for the out-of-band region. The bandwidth of such systems can also be defined as more than 25% of a center frequency, or more than 1.5 GHz. Clearly, this bandwidth is much greater than the bandwidth used by any current technology for communication. The large bandwidth of UWB signals provides robustness to jamming, and has properties with a low probability of detection. UWB devices usually require low transmit power, due to control over duty cycle, thus allowing a longer battery life for handheld equipment. In addition to its high data rate and short-range features that make UWB a suitable candidate for the wireless body-centric network, the low power level also increases the compatibility of this technology with currently existing wireless systems.

For narrowband body-centric wireless systems, comparison of the results taken in the anechoic chamber to those in the laboratory shows a noticeable increase in signal variability outside the anechoic chamber, which is due to multipath fading. Outside the chamber, the propagation channel is not stable, even when the body is still, which is caused by people moving within the room. Preliminary frequency-domain measurements indicate that UWB on-body, like its narrowband counterpart, suffer significant radio link variation due to different body postures. However, it is noted that the human body attenuates the UWB signal (3 to 9 GHz) more severely, so that at NLOS situations, the received signal level becomes very low. The results also indicate that the dominant propagation channel for UWB on-body networks is still the free-space path, although it depends on the type of antenna used.

10.1.3 Overview of Applications

Body-centric wireless communications is aimed at providing systems with constant availability, reconfigurability, and unobtrusiveness. High levels of processing and complex network protocols are needed to provide the powerful computational functionalities required for advanced applications. These requirements have led to increased research and development activities, with the main interests being health care and patient monitoring, personal identification, navigation, personal multimedia entertainment, and task-specific/fully compatible wireless wearable computers. Body-centric wireless communications can be applied to many fields, and a summary of their applications can be found in Table 10.2. Some of them include:

- *Medical applications:* smart diagnosis, treatment and drug delivery system, patient monitoring, and aging care;
- *Wireless access/identification systems:* wireless transactions and identification of individual peripheral devices;
- *Navigation support and location based services:* tourism, security, and intelligent transportation system;

Table 10.2   Summary of Potential Applications in Body-Centric Wireless Communications

Standard	Applications Scenarios	Wave Propagation Mechanism	Chapter Number in the Book
UHF/VHF	Medical implants, personal access, personal identification system	Inductive and near-field coupling	4, 9
Medical implant (licensed)	Medical implants, telemedicine	Near-field, far-field, lossy media	7, 8, 9
BodyLAN	Telemedicine	Free-space wave	7
Bluetooth/ ZigBee/WLAN	Personal multimedia entertainment, security, smart clothing, location-based services	Space and surface waves	3, 6, 8
UWB	Wireless DVD, wearable and ubiquitous computing	Space and surface waves	5

- *Personal multimedia entertainment:* wireless DVD and wearable computing;
- *Military and space applications:* smart suits, battlefield personnel care and intelligence, and biosensors for astronaut monitoring.

10.1.4   Conclusions

Body-centric wireless communications consist of a number of nodes and units placed on the human body or in close proximity to it. They are seen to be a central component in the fourth generation communication systems, as envisaged, for example, in the EC MAGNET Integrated Project [8], that will embrace both local and global connectivity, and include body area networks (BANs), personal area networks (PANs), and local area networks (LANs). Currently, body-centric networks are used to pass simple information that requires very low processing capabilities. However, some high-performance units will be needed in the future to provide the facilities for powerful computational processing with high data rates for applications, such as video streaming and fast data communications.

The properties of in-body and on-body wireless links are very different from other forms of communications (e.g., personal mobile systems), in that they are very dependent on both the material properties, the design of the antennas, and the physical properties of the body. For communications on the body, research in this book indicates that by the use of appropriate antennas, it is possible for the body to act as a waveguiding structure, confining most of the radio energy within a wavelength of the surface. This is very spectrally efficient, since there is low leakage of radiation outside of the body zone.

10.2   Future Challenges

10.2.1   Towards Air Interfaces and New Standards

Existing technologies to connect portable devices are based on personal area network standards, such as 802.11x or Bluetooth. Neither of these are spectrum-efficient for body area networks, since most of the radio energy is directed away from the body when the radio antenna is placed close to the skin. In addition, only the ISM band at 2.4 GHz is conventionally suitable for this application, since body absorption at 5 GHz (or at UWB frequencies) is generally considered too high to allow this frequency to be used. As the 2.4-GHz band is now widely used for WLAN applications, body area networks using this band are likely to become sources of mobile interference, thus further degrading existing WLAN performance.

Since on-body communications is an area in which standards are only just starting to be defined, this is an ideal time to influence the debate on technologies. Market forces on their own are unlikely to develop the most spectrum-efficient solution, since these systems use license-exempt spectrum that is considered as a free resource by the portable device manufacturers. It is therefore our belief that now is the optimum time to consider forward-looking, precompetitive research into how body-centric wireless networks (mainly on-body and off-body) should be implemented in a spectrally efficient manner without inconveniencing other radio users.

10.2.2 Novel Antennas for Body-Centric Wireless Communications

For on-body or in-body environments, antennas are required to conform to the body, and to be immune from frequency and polarization detuning. It is also likely that the associated electronics and sensors will be conformal in shape. While many antennas in wearable equipment will be in conventional miniaturized form, other antennas will be made using textiles, particularly for specialist occupations such as firefighters and paramedics, and where fashion dictates form and design. Antennas for medical implants are particularly challenging, due to a very variable environment and the need for low-frequency operation using small antennas. In antenna design, physical dimensions, directivity, efficiency, and so forth, are constrained by the wavelength of the radiation. These constraints are very relevant to body-centric antenna applications.

For on-body communication systems, it is vital to understand how best to specify an antenna radiation pattern, when part of the radiation is space and part in the lossy body; and how to specify coupling into the propagation mode, which may be a surface wave or free-space wave, or a combination of both. Such work would be based on efficient numerical simulation plus verification by phantoms. Thus, it is essential to develop special antennas for on-body communication networks, which need monopole patterns for coupling to surface wave and patchlike patterns for surface/space wave links. Such systems also need to minimize SAR and off-body radiation, unless they are also used as wearable antennas for PANs.

To minimize multipath reflections from various body parts for a specific on-body communication link, investigation of switchable and diversity antennas are needed to overcome fading. This will require studies of multiple antennas on the body, both in correlation and in isolation from each other. This could also be extended to include MIMO [9] performance on body.

10.2.3 Medical Implant Systems

The use of medical implants will increase as the number and capability of nanosensor and microsensor devices increase. Communications between implants may then be used to form networks, which improve the system functionality. Figure 1.5 shows a possible concept for a hierarchical network of sensors in the brain. There are significant challenges in such networks. First, there is a drive to reduce the size of the whole implant to the same size as the sensor. This implies that small antennas will be working in a lossy, variable medium. Second, the antennas may also have to function as receptors in inductive power transfer systems. The communications may use the same frequency or a higher frequency, depending on the required data rate. Real-time monitoring of many data channels from each sensor, monitoring electrical activity at a high resolution in the brain, for example, will require much higher data rates than have currently been achieved. Of course, power is at a premium in implants, so the power transfer system is as crucial as the communications. Optimum design of power transfer and communications relies on a good understanding of the electromagnetics of antennas and propagation. System design may be unique to the implant function required, but for future design and production efficiency through commonality, more systematic studies of antennas and propagation for implants are required.

Although there has been significant progress in simulation tools, there is still a need for improvement, particularly in run time for full-body simulators. It is clear from Chapter 9 that there is considerable variability of the propagation channel during body movement, and that the channel varies between different body types, particularly between adult and child. More work to characterize the channels for implants is required. Similarly, there are great difficulties in the measurement of implant antennas and channels on living bodies, and not much data exists. Phantom measurements are possible but time-consuming, and the materials require careful calibration. Improvements in the benchmarking of simulations, through measured data, are needed.

10.2.4 Characterization of Time-Domain Systems

The best choice of measurement techniques for the characterization of time-domain systems for UWB communications is still unclear. Measurements can be made in either the spectral domain or the time domain. Conventionally, path loss measurements are based on the spectral domain. However, to fully characterize a UWB channel on the body, fast, accurate, and full-bandwidth measurements are needed. One problem is that the human body is variable over the duration of frequency sweep, particularly during fast movements. This means that it is impossible to properly record communication channel variation over a wide bandwidth. The on-body UWB signals should probably be most accurately measured in the time domain. Here, the equipment mobility will be very important, since UWB on-body channel measurement is thought to be highly lossy, frequency-dependent, and with strong multipath effects. UWB pulse distortion due to the channel dispersion may well be also body temperature–dependent, and measurement campaigns should be carefully prepared.

References

[1]   http://www.fitsense.com/b/BodyLAN.asp.

[2]   Official Bluetooth Wireless Information Web site, http://www.bluetooth.com/bluetooth/.

[3]   Cook, D. J., and S. K. Das, (eds.), *Smart Environments: Technologies, Protocols, and Applications*, New York: John Wiley & Sons, 2004.

[4]   IEEE 802.15 Working Group for WPAN, http://www.ieee802.org/15/.

[5]   IEEE 802.11 Working Group for WLAN, http://grouper.ieee.org/groups/802/11/.

[6]   FCC, The Medical Implant Communications Service (MICS), http://wireless.fcc.gov/services/personal/medicalimplant/.

[7]   FCC First Report and Order, Revision of the Part 15 Commission's Rules Regarding Ultra-Wideband Transmission Systems, ET-Docket 98-153, April 22, 2002.

[8]   The European IST-507102, "My Personal Adaptive Global NET (MAGNET)," http://www.ist-magnet.org.

[9]   Chuah, C. N., et al., "Capacity Scaling in MIMO Wireless Systems Under Correlated Fading," *IEEE Trans. on Information Theory*, Vol. 48, No. 3, March 2002, pp. 637–650.

About the Authors

Akram Alomainy received an M.Eng. in communication engineering from Queen Mary, University of London (QMUL), United Kingdom, in July 2003. In September 2003, he commenced his research studies in the Electronic Engineering Department, QMUL, where he is currently working toward his Ph.D. His current research interests include small and compact antennas for wireless body area networks, radio propagation characterization and modeling for body-centric networks, antenna interactions with the human body, and computational electromagnetic and advanced antenna enhancement techniques. Mr. Alomainy is a member of the IEE and the IEEE. He has attended a number of established international conferences and workshops, and has many conference and journal publications.

Costas Constantinou received a B.Eng. (Hons.) in electronic and communications engineering and a Ph.D. in electronic and electrical engineering from the University of Birmingham, United Kingdom, in 1987 and 1991, respectively. In 1989, he joined the School of Electronic and Electrical Engineering at the University of Birmingham as a lecturer and subsequently as a senior lecturer in the Communications Engineering Research Group. His research interests span many aspects of modern communications engineering—electromagnetic theory, electromagnetic scattering and diffraction, electromagnetic measurement, radiowave propagation modeling, mobile radio, wireless networks, future adaptive communication network architectures, and the modeling of very large-scale networks using methods of statistical physics.

Philippe De Doncker received a physics engineering degree in 1996 and a Ph.D. in 2001, both from the Université Libre de Bruxelles, Belgium, where he is now an assistant professor. His research focus is on propagation and statistical methods for wireless communications.

Patrick Claus Friedrich Eggers since 1992 has been the project leader of the propagation group of the Center for PersonKommunikation (CPK) at Aalborg University (AAU), where he is an associate professor. He is now the research coordinator of the Antennas and Propagation Division, Department of Communication Technology (KOM), AAU. He is on the technical research council of the Center for TeleInFrastruktur (CTIF) at AAU and has been project manager and work package manager in several European research projects (e.g., TSUNAMI, CELLO, and so forth) and in industrial projects with partners such as Nokia, Ericsson, Motorola, IOSpan, ArrayComm, Avendo Wireless, Samsung, and so forth, and as a representative of CTIF, KOM, and previously the CPK. He is the author of more than 40 papers, as well as the section author and chapter editor in different COST final reports (COST207, 231, 259) and books on ultrawideband and cooperative communication. He is the initiator and coordinator of an internationally targeted M.Sc.E.E. program in mobile communications, taught in English at Aalborg University, as well as designer and coordinator of the newly starting program in software defined radio.

Noel E. Evans received a B.Sc. in 1973 and an M.Sc. in 1974 from Queen's University of Belfast, Northern Ireland, United Kingdom. He obtained a Ph.D. from Queen's University in 1977 for work on programmable transversal filters built using intracell charge-coupled devices. Following further research into narrowgate CCDs at Queen's, he taught analog electronics and communications at the University of Ulster, from 1980 to 2006. Here, he also developed research programs into biomedical instrumentation, human and animal physiological signal acquisition using radio frequency techniques, and electromagnetic wave propagation at frequencies extending across the HF-UHF bands. He now works as an independent consultant in the field of RF devices.

Andrew Fort received a bachelor's degree in computer engineering from the University of Victoria, Canada, in 1998. As a part of his cooperative study program, he worked at Sanyo Electric, Japan, developing hardware for digital cellular phones. After graduation, he worked at IVL Technologies, Victoria, designing DSP algorithms for pitch recognition of the human voice. He joined IMEC, Belgium, in January 2000, and was involved in the research and development of a wide range of communication systems, including satellite, multiple antenna, and wireless local area networks. Since January 2004, he has conducted doctoral studies at the Vrije Universiteit Brussel on ultralowpower medical sensors and body area communication networks, under the supervision of Professor Leo Van Biesen.

Flemming Bjerge Frederiksen received an M.Sc.E.E. from Aalborg University, Denmark, in 1981. From July 1981 to May 1988, he was employed by Aalborg University, as teacher and research engineer and subsequently as assistant professor. His research was focused on the new possibilities within analog and digital signal processing (e.g., switched capacitor techniques). Since June 1988, he has been employed as an associate professor by the Institute of Electronic Systems, Aalborg University. From February 1989 to March 1993, he was the head of the Study Board (Head of Studies), with responsibility for the Engineering Studies on Electronic Engineering and on Computer Engineering at Aalborg University (B.Sc.E.E. and M.Sc.E.E.). In April 1993, he was elected the Head of Institute for the Institute of Electronic Systems, Aalborg University, and he is currently appointed for this position. Presently, the institute covers the following scientific areas: communication technology, acoustics, and control engineering. Today, his research work is within orthogonal frequency division multiplexing (OFDM) and other techniques towards improvements on the capacity of digital wireless communications. He is a senior member of the IEEE and a member of the Danish Engineering Society (IDA).

Katsuyuki Fujii received a B.E. and an M.E. in electronic engineering from Chiba University, Chiba, Japan, in 2001 and 2003, respectively. He is currently a Ph.D. candidate at Chiba University. His main interest is the evaluation of the interaction between electromagnetic fields and the human body by use of numerical and experimental phantoms. He received the Technical Achievement Award from the Institute of Image Information and Television Engineers of Japan (ITE) in 2002, the Suzuki Memorial Achievement Award from the ITE in 2004, the 2004 International Symposium on Antennas and Propagation Poster Presentation Award from the IEICE, 2005 IEEE Tokyo Student Workshop IEICE Best English Prize from the IEICE Student Branch Doctoral Student Forum, 2005 Young Researcher's Award from the IEICE, and 2005 The 2nd IEEE Tokyo Student Workshop First Prize. Mr. Fujii is a member of the IEEE and the ITE and a student member of the IEICE.

Peter S. Hall is a professor of communications engineering and the head of the Communications Engineering Research Group in the School of Electronic Electrical and Computer Engineering at the University of Birmingham. After graduating with a Ph.D. in antenna measurements from Sheffield University, he spent three years with Marconi Space and Defence Systems, Stanmore, working largely on a European communications satellite project. He then joined the Royal Military College of Science as a Senior Research Scientist, progressing to Reader in Electromagnetics. He joined the University of Birmingham in 1994. He has researched extensively in the areas of microwave antennas and associated components and antenna measurements. He has published four books, more than 230 learned papers, and has taken various patents. These publications have earned six IEE premium awards, including the 1990 IEE Rayleigh Book Award for the *Handbook of Microstrip Antennas*. Professor Hall is a fellow of the IEEE and the IEE. He is a former chairman of the IEE Antennas and Propagation Professional Group and is an IEEE Distinguished Lecturer and a member of the IEEE Fellow Application Review Board. He chaired the organizing committee of the 1997 IEE International Conference on Antennas and Propagation, and has been associated with the organization of many other international conferences. He was honorary editor of *IEE Proceedings Part H Journal* from 1991 to 1995 and is currently on the editorial boards of the *International Journal of RF and Microwave Computer Aided Engineering* and *Microwave and Optical Technology Letters*.

Yang Hao received a Ph.D. from the Centre for Communications Research (CCR) at the University of Bristol, United Kingdom, in 1998. From 1998 to 2000, he was a postdoctoral research fellow at the School of Electrical and Electronic Engineering, University of Birmingham, United Kingdom. In May 2000, he joined the Antenna Engineering Group, Queen Mary College, University of London, London, United Kingdom, first as a lecturer and then as a reader in antenna and electromagnetics. Dr. Hao is active in computational electromagnetics, electromagnetic bandgap structures and microwave metamaterials, antennas and radio propagation for body-centric wireless networks, active antennas for millimeter/submillimeter applications, and photonic integrated antennas. He has published more than 70 technical papers, and has served as an invited and keynote speaker, a conference organizer, and a session chairperson at many international conferences. Dr. Hao is a senior member of the IEEE and a member of the Technical Advisory Panel of IEE Antennas and Propagation Professional Network.

Koichi Ito received a B.S. and an M.S. from Chiba University, Chiba, Japan, in 1974 and 1976, respectively, and a D.E. from the Tokyo Institute of Technology, Tokyo, Japan, in 1985, all in electrical engineering. From 1976 to 1979, he was a research associate with the Tokyo Institute of Technology. From 1979 to 1989, he was a research associate with Chiba University. From 1989 to 1997, he was an associate professor with the Department of Electrical and Electronics Engineering, Chiba University, and is currently a professor with the Research Center for Frontier Medical Engineering, as well as with the Faculty of Engineering, Chiba University. He has been appointed as one of the deputy vice-presidents for research, as well as the director of the Office of Research Administration, Chiba University. In 1989, 1994, and 1998, he was with the University of Rennes I, France, as an invited professor. His main research interests include analysis and design of printed and small antennas for mobile communications, research on evaluation of the interaction between electromagnetic fields and the human body by use of numerical and experimental phantoms, and microwave antennas for medical applications such as cancer treatment. Dr. Ito is a member of the AAAS, the IEICE of Japan, the Institute of Image Information and Television Engineers of Japan (ITE), and the Japanese Society of Hyperthermic Oncology. He served as the chairperson of Technical Group on Radio and Optical Transmissions of ITE from 1997 to 2001. He also served as chairperson of the IEEE AP-S Japan Chapter from 2001 to 2002. He is the chairperson of the Technical Group on Human Phantoms for Electromagnetics, IEICE, the vice-chairperson of the 2007 International Symposium on Antennas and Propagation (ISAP2007), and an associate editor of the *IEEE Transactions on Antennas and Propagation*.

Anders J. Johansson received a master's degree, a Lic. Eng., and a Ph.D. in electrical engineering from Lund University, Lund, Sweden, in 1993, 2000, and 2004, respectively. From 1994 to 1997, he was with Ericsson Mobile Communications AB, developing transceivers and antennas for mobile phones. Since 2005, he has been an associate professor at the Department of Electroscience at Lund University. His research interests include antennas and wave propagation for medical implants, as well as antenna systems and propagation modeling for MIMO systems. He is responsible for both the antenna measurement laboratory at the Department of Electroscience, and the circuit theory education at the Computer Science and Engineering Masters program.

Muhammad Ramlee Kamarudin obtained his first degree from Universiti Teknologi Malaysia (UTM), Malaysia, with Honours, majoring in electrical and telecommunication engineering, graduating in 2003. He was then a lecturer at the Faculty of Electrical (FKE), UTM. He has been awarded a scholarship from the government of Malaysia to further study in the United Kingdom. He received an M.Sc. in communication engineering at the University of Birmingham in 2005. Currently, he is working towards a Ph.D., with his thesis entitled, "Antennas for On-Body Communications," at the University of Birmingham, supervised by Professor Peter S. Hall.

Anders Karlsson received an M.Sc. and a Ph.D. from Chalmers University of Technology, Gothenburg, Sweden, in 1979 and 1984, respectively. Since 2000, he has been a professor at the Department of Electroscience at Lund University, Lund, Sweden. His research activities include scattering and propagation of waves, inverse problems, and time-domain methods. Currently, he is involved in projects concerning propagation of light in blood, wireless communication with implants, and design of passive components on silicon.

Hiroki Kawai received a B.E., an M.E., and a D.E., all in electrical engineering from Chiba University, Chiba, Japan, in 1999, 2001, and 2005, respectively. He is currently a researcher at the National Institute of Information and Communications Technology (NICT), Japan. His main interests include analysis and design of small antennas for mobile communications, and research on evaluation of the interaction between electromagnetic fields and the human body, using numerical and experimental phantoms. He is author or coauthor of more than 40 research papers in reference journals and conferences. Dr. Kawai is a member of the Institute of Electrical, Information and Communication Engineers (IEICE), Japan, the IEEE, and the Bioelectromagnetics Society (BEMS).

Maciej Klemm received an M.Sc. in microwave engineering from Gdansk University of Technology, Poland, in 2002. In February 2003, he joined the Electronics Laboratory, Swiss Federal Institute of Technology (ETH), Zurich, Switzerland, where he earned a Ph.D. in 2006. Currently, he is a postdoctoral researcher at the University of Bristol, United Kingdom. His current research interests focus on the development of UWB radar for breast cancer detection, small UWB antennas and UWB communications, antenna interactions with a human body, and electromagnetic simulations. He holds an M.Sc., a Ph.D., and an MIEEE.

István Z. Kovács received a B.Sc. from "Politehnica" Technical University of Timisoara, Romania, in 1989; an M.Sc.E.E. from the Franco-Polish School of New Information and Communication Technologies/Ecole Nationale Supérieure des Télécommunications de Bretagne, Poland/France, in 1996; and a Ph.D.E.E. in wireless communications from Aalborg University, Denmark, in 2002. From 2002 to 2005, he held the position of assistant research professor with the Center for TeleInFrastruktur, Department of Communication Technology of Aalborg University, in the antennas and propagation division. His research interests were in the field of radio channel propagation measurements and modeling, with major focus on short-range ultrawideband radio channel and ultrawideband antenna investigations. He was actively involved in the European IST PACWOMAN and IST MAGNET projects, and participated in several industrial projects with partners such as TeleDanmark, Motorola, IOSpan, and ArrayComm. He has written a number of papers and has contributed chapters to books on UWB propagation topics. Currently, he has the position of wireless networks specialist in network systems research, Nokia Networks, Aalborg, Denmark, conducting research in the area of 3G/3.9G wireless networks.

Yuriy I. Nechayev graduated from Kharkiv State University (Ukraine) with a Diploma of Specialist in physics in 1996. He received a Ph.D. in electronic and electrical engineering from the University of Birmingham, United Kingdom, in 2005. His Ph.D. thesis title was "Investigations into Propagation Mechanisms for Urban Radiowave Propagation Modelling." Since 2003, he has been with the University of Birmingham as a research associate, and later as a research fellow, carrying out research on radiowave propagation around the human body. His research interests include electromagnetics, radiowave propagation, diffraction and scattering, and propagation in random media.

Clive Parini is the head of both the Communications Research Group and the Antenna & Electromagnetics Laboratory at Queen Mary, where there is a strong emphasis on electromagnetic modeling using a variety of methods, including physical optics, geometric optics, GTD/UTD, method of moments, modal matching, FDTD, and numerical optimization. Professor Parini has published more than 150 papers on research topics, including array mutual coupling, array beamforming, antenna metrology, microstrip antennas, millimeter-wave compact antenna test ranges, and millimeter-wave integrated antennas. In January 1990, he was one of three coworkers to receive the IEE Measurements Prize for work on near-field reflector metrology. In 2002, along with several coworkers, he was awarded the BAE SYSTEMS Chairman's Bronze Award for Innovation for work on microwave near-field metrology. He is a fellow of the IEE and recently stepped down as the chairman of the IEE Antennas & Propagation Professional Network Executive Team. He is currently an honorary editor for the *IEE Journal of Microwaves, Antennas and Propagation*. He is a founder of Agent-Tel, a Queen Mary–owned telecommunication start-up company, exploiting agent technology and smart antennas to optimize capacity in mobile cellular networks. He has recently taken up a part-time post as the director of the Science and Engineering Graduate School at Queen Mary.

Gert Frølund Pedersen received a B.Sc. E.E., with honors, in electrical engineering from the College of Technology in Dublin, Ireland, and an M.Sc. E.E. and a Ph.D. from Aalborg University, in 1993 and 2003, respectively. He has been employed since 1993 by Aalborg University, where he currently is working as a professor for the Antenna & Propagation Group. His research has focused on radio communication for mobile terminals, including small antennas, antenna systems, propagation, and biological effects. He has also worked as a consultant for developments of antennas for mobile terminals, including the first internal antenna for mobile phones in 1994 with very low SAR, the first internal triple-band antenna in 1998 with low SAR and high efficiency, and various other antenna diversity systems that are rated as the most efficient on the market. Recently, he has been involved in establishing a method to measure the communication performance for mobile terminals, which can be used as a basis for a 3G standard in which measurements including the antenna will be needed. He is also involved in small terminals for 4G, including several antennas (MIMO systems) and ultrawideband antennas to enhance data communication.

Ramjee Prasad is a distinguished educator and researcher in the field of wireless networks, packet communications, multiple access protocols, access techniques, and multimedia communications. Currently, he is a professor and the director of the Center for TeleInFrastruktur (CTIF) at Aalborg University, Denmark. He is the project coordinator of the European 6th framework integrated project, "My Personal Adaptive Global Net Beyond (MAGNET Beyond)." He has published more than 500 technical papers, contributed to several books, and has authored, coauthored, and edited 19 books. Professor Prasad is the editor-in-chief of the *Springer International Journal on Wireless Personal Communications*. He was the founding chairman of the European Centre of Excellence in Telecommunications, known as HERMES. He now continues to contribute as an honorary chairman of HERMES. He is a fellow of the IEE, a fellow of the IETE, a member of NERG, and a senior member of the IEEE. Professor Prasad is the recipient of several international awards. He recently received the Telenor 2005 Nordic Research Award.

Yahya Rahmat-Samii is a distinguished professor of the Electrical Engineering Department at the University of California, Los Angeles (UCLA). Before joining UCLA, he was a senior research scientist at the NASA Jet Propulsion Laboratory (JPL). He became a fellow of the IEEE in 1985, and was elected as the president of IEEE Antennas and Propagation Society (AP-S) in 1995. Professor Rahmat-Samii has published more than 650 journal and conference papers, and more than 20 books or book chapters in the areas of electromagnetics and antennas. He has had pioneering research contributions in diverse areas of electromagnetics, antennas, measurement and diagnostics techniques, numerical and asymptotic methods, satellite and personal communications, human/antenna interactions, frequency selective surfaces, electromagnetic bandgap structures, applications of the genetic algorithms, and particle swarm optimization, and so forth (see http://www.ee.ucla.edu/antlab). In 1992 and 1995, he was the recipient of the Best Application Paper Prize Award (Wheeler Award), for papers published in *IEEE AP-S Transactions*. In 1999, he was the recipient of the University of Illinois ECE Distinguished Alumni Award. In 2000, Professor Rahmat-Samii received the IEEE Third Millennium Medal and the AMTA Distinguished Achievement Award. In 2001, he received an honorary doctorate in physics from one of the oldest universities in Europe, the University of Santiago de Compostela, Spain. In 2001, he was elected as the foreign member of the Royal Academy of Belgium for Science and the Arts. In 2002, he received the Technical Excellence Award from the JPL. He is the recipient of the 2005 International Union of Radio Science (URSI) Booker Gold Medal, presented at the URSI General Assembly, New Delhi, India. Professor Rahmat-Samii is also the designer of the IEEE AP-S logo.

Tareq Salim received a B.Sc. in communication and electronic engineering from Applied Science University, Jordan, in 2002, and an M.Sc. in communication engineering systems from the University of Birmingham, United Kingdom, in 2003. He is currently working to finish his Ph.D. at the University of Birmingham. His research interests include wave propagation in layered medium with singularities, numerical analysis and integration, wave scattering, inverse scattering problems and statistical EM, applied numerical methods for EM (FDTD and FEM), parallel computation methods with FDTD-MPI or mass-matrix FEM, hybrid multipole numerical method with uniform theory of geometrical diffraction method (UTD), antenna measurements and design, genetic algorithms, object-oriented programming, theoretical and analytical methods in EM, and EBG materials.

Pekka Salonen received an M.Sc. (Tech.) and a Dr. (Tech.) from the Tampere University of Technology (TUT), Tampere, Finland, in 1997 and 2001, respectively, both in electrical engineering. He was a research doctor with the Institute of Electronics, TUT, from 2002 to 2003, and a visiting postdoctoral researcher with ARAM Laboratory, University of California, Los Angeles, from 2003 to 2004, under the supervision of Professor Yahya Rahmat-Samii. Dr. Salonen is currently with Patria's Systems business unit. His research interests are in the areas of wearable antennas, antennas for mini-UAV systems, antennas for maritime applications, and computational electromagnetics.

William G. Scanlon received a first-class honors B.Eng. engineering degree through part-time study at the University of Ulster, United Kingdom, in 1994. In 1997, he was awarded a Ph.D. from the University of Ulster, for work on the numerical modeling of antenna-body interactions at radio frequencies, and has been employed as an academic since then. He is currently a senior lecturer in telecommunications at Queen's University, Belfast, United Kingdom. Dr. Scanlon's current research interests include antennas, radiowave propagation, bioelectromagnetics, wireless networks, and link-layer protocols. He has almost 10 years of experience within the electrical and electronic industries, having worked for Nortel Networks as a senior RF engineer, and with Siemens UK as a project engineer.

Rainee N. Simons received a B.S. in electronics and communications engineering from Mysore University, Mysore, India, in 1972; an M.Tech. in electronics and communications engineering from the Indian Institute of Technology (IIT), Kharagpur, India, in 1974; and a Ph.D. in electrical engineering from the IIT, New Delhi, India, in 1983. In 1979, he began his career as a senior scientific officer with the IIT, where he was involved with finline components for millimeter-wave applications, and X-band toroidal latching ferrite phase shifters for phased arrays. Since 1985, he has been with the Communications Technology Division (CTD), NASA Glenn Research Center (GRC), Cleveland, Ohio, where he is currently the chief of the Electron Device Technology Branch. While with NASA GRC, he was involved with optical control of MESFET and high electron mobility transistor (HEMT) devices, high-temperature superconductivity, modeling of coplanar waveguide/stripline (CPW/CPS) discontinuities, CPW feed systems for printed antennas, linearly tapered slot arrays, packaging of monolithic microwave integrated circuits (MMICs), silicon carbide (SiC) diode mixers, RF microelectromechanical systems (MEMS), and GaN solid-state power amplifiers. He is currently involved in the area of high-efficiency space traveling wave tube (TWT) power combiners for communications from near-Earth as well as deep space. He has authored or coauthored more than 130 publications in refereed journals and international symposium proceedings. He is the author of *Optical Control of Microwave Devices* (Artech House, 1990). He also authored the chapter "High Temperature Superconducting

Coplanar Waveguide Microwave Circuits and Antennas," which appeared in *Advances in High-Tc Superconductors* (Trans-Tech Publications, 1993). He coauthored the chapter "Tapered Slot Antenna," which has appeared in *Advances in Microstrip and Printed Antennas* (John Wiley & Sons, 1997). He is the author of *Coplanar Waveguide Circuits, Components, and Systems* (John Wiley & Sons, 2001). He holds three U.S. patents. Dr. Simons has organized workshops, chaired sessions, and served on the technical program committees of several IEEE international symposiums. He is a member of the editorial board of the *IEEE Transactions on Microwave Theory and Techniques* and is an associate editor of the *IEEE Transactions on Antennas and Propagation*. He was the recipient of the Distinguished Alumni Award. He is a recipient of more than 17 NASA Tech Brief Awards. He was also a recipient of the NASA Public Service Medal in 2001, for pioneering research and development of microwave printed antennas and distribution media. He is a fellow of the IEEE.

Masaharu Takahashi received a B.S. in electrical engineering from Tohoku University, Tohoku, Japan, in 1989; and an M.S. and a Ph.D. from the Tokyo Institute of Technology, Tokyo, Japan, in 1991 and 1994, respectively. From 1994 to 1996, he was a research associate; from 1996 to 2000, he was an assistant professor at Musashi Institute of Technology, Tokyo, Japan; and from 2000 to 2004, he was an associate professor at Tokyo University of Agriculture and Technology, Tokyo, Japan. He is currently an associate professor at Chiba University, Chiba, Japan. His research interests include electrically small antennas, planar array antennas, evaluation of the interaction between electromagnetic fields and the human body by use of numerical and experimental phantoms, and electromagnetic compatibility. Dr. Takahashi was the recipient of the 1994 IEEE Antennas and Propagation Society Tokyo Chapter Young Engineer Award. He is a senior member of the IEEE.

Index

For further information on these and other Artech House titles,

including previously considered out-of-print books now available through our In-Print-Forever® (IPF®) program, contact:

Artech House
685 Canton Street
Norwood, MA 02062
Phone: 781-769-9750
Fax: 781-769-6334
e-mail: artech@artechhouse.com

Artech House
46 Gillingham Street
London SW1V 1AH UK
Phone: +44 (0)20 7596-8750
Fax: +44 (0)20 7630 0166
e-mail: artech-uk@artechhouse.com

Find us on the World Wide Web at:
www.artechhouse.com